T0073043

Sustainable Shale Oil and Gas

Emerging Issues in Analytical Chemistry

Series Editor
Brian F. Thomas

ELSEVIER

AMSTERDAM • BOSTON • HEIDELBERG • LONDON
NEW YORK • OXFORD • PARIS • SAN DIEGO
SAN FRANCISCO • SINGAPORE • SYDNEY • TOKYO

Sustainable Shale Oil and Gas

Analytical Chemistry, Geochemistry, and Biochemistry Methods

Vikram Rao
Research Triangle Energy Consortium, Research Triangle Park, NC, United States

Rob Knight
University of California San Diego, La Jolla, CA, United States

AMSTERDAM • BOSTON • HEIDELBERG • LONDON
NEW YORK • OXFORD • PARIS • SAN DIEGO
SAN FRANCISCO • SINGAPORE • SYDNEY • TOKYO

Elsevier
Radarweg 29, PO Box 211, 1000 AE Amsterdam, Netherlands
The Boulevard, Langford Lane, Kidlington, Oxford OX5 1GB, United Kingdom
50 Hampshire Street, 5th Floor, Cambridge, MA 02139, United States

Published in cooperation with RTI Press at RTI International, an independent, nonprofit research institute
that provides research, development, and technical services to government and commercial clients worldwide
(www.rti.org). RTI Press is RTI's open-access, peer-reviewed publishing channel. RTI International is a trade
name of Research Triangle Institute.

Notices

Knowledge and best practice in this field are constantly changing. As new research and experience broaden
our understanding, changes in research methods, professional practices, or medical treatment may become
necessary.

Practitioners and researchers must always rely on their own experience and knowledge in evaluating and using
any information, methods, compounds, or experiments described herein. In using such information or methods
they should be mindful of their own safety and the safety of others, including parties for whom they have a
professional responsibility.

To the fullest extent of the law, neither the Publisher nor the authors, contributors, or editors, assume any
liability for any injury and/or damage to persons or property as a matter of products liability, negligence or
otherwise, or from any use or operation of any methods, products, instructions, or ideas contained in the
material herein.

ISBN: 978-0-12-810389-0

British Library Cataloguing-in-Publication Data
A catalogue record for this book is available from the British Library

Library of Congress Cataloging-in-Publication Data
A catalog record for this book is available from the Library of Congress

For Information on all Elsevier publications visit
our website at https://www.elsevier.com

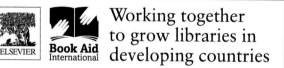

Working together
to grow libraries in
developing countries

ELSEVIER Book Aid International

www.elsevier.com • www.bookaid.org

Publisher: John Fedor
Acquisition Editor: Kathryn Morrissey
Editorial Project Manager: Jill Cetel
Production Project Manager: Vijayaraj Purushothaman
Designer: Matthew Limbert

Typeset by MPS Limited, Chennai, India

DEDICATION

To *Campastimes*, the Indian Institute of Technology Madras magazine, sadly now defunct, where I cut my baby teeth in writing 54 years ago.

CONTENTS

LIST OF CONTRIBUTORS

Jason J. Amsden
Nanomaterials and Thin Films Laboratory, Department of Electrical and Computer Engineering, Duke University, Durham, NC, United States

Rob Knight
Pediatrics and Computer Science & Engineering, University of California San Diego, La Jolla, CA, United States

Tim Profeta
Nicholas Institute for Environmental Policy Solutions, Duke University, Durham, NC, United States

Vikram Rao
Research Triangle Energy Consortium, Research Triangle Park, NC, United States

Nicole M. Scott
Biota Technology, San Diego, CA, United States

Luke K. Ursell
Biota Technology, San Diego, CA, United States

David S. Vinson
Department of Geography and Earth Sciences, University of North Carolina at Charlotte, Charlotte, NC, United States

Joel Walls
Ingrain Inc., Houston, TX, United States

The ability of the United States to utilize domestic unconventional oil and gas is the energy issue of the day. Cheap energy from these sources is driving manufacturing growth in the country and suppressing fuel costs throughout the economy, the utility sector is moving rapidly from coal generation to natural gas, and the carbon dioxide emissions from the utility sector have dropped to their lowest since the 1990s. A fierce debate, however, has arisen over whether this transition is a net good for the environment. In particular, many question whether the climate, air, and water impacts from acquiring oil and gas through unconventional processes overshadow any benefits from the use of a cleaner fuel.

The ferocity of this political debate has clearly outpaced policymakers' understanding of the problem, as well as the assessment of the potential solutions. Also, the speed of technological development continues to accelerate past the government's knowledge of how to monitor and assess the problem. We thus are left with a policy challenge laced with emotion but lacking the information needed to make wise policy.

Wise environmental policy cannot be constructed under this circumstance. To properly evaluate and weigh the options, policymakers need clear-eyed assessments of both the scale of the issue and the potential solutions to the risks.

In this book, Rao and Knight succinctly summarize the state of knowledge on the management of the environmental risks of accessing shale oil and gas. By standing on shoulders of the investments of entities such as the US Department of Energy and the Environmental Defense Fund, and also the research of many experts in the academy (including many of my colleagues at Duke University), the authors seek to create a library of knowledge that should be the "raw material" for wise policy. They also explore new measurement technologies by which we might inform regulation.

Energy is the lifeblood of the economy—and natural gas is a game changer in that it may provide that blood to our economy for years to come. Unconventional natural gas, on the other hand, could prove to be too environmentally challenging to make it a longer term solution to our energy needs. Which future we pursue, in the end, will depend on the issues explored in this book.

Tim Profeta

PREFACE

The last sentence of this book reads, "That which cannot be measured, cannot be regulated or otherwise controlled or exploited." This embodies the essential motivation for the book. To put that in perspective, I quote the last lines from a previous book of mine:

> Low-cost energy is a tide that lifts all boats of economic growth. Shale gas is a powerful such tide. It has burst upon us so unexpectedly that we have become rattled by the flotsam it carried with it. This author concludes that the flotsam is manageable, allowing us to enjoy the benefits of the tide.

By flotsam, I meant the environmental baggage. Although I still subscribe to this conclusion, I have increasingly come to believe that current measurement techniques are inadequate for the special circumstances surrounding this resource, which now importantly includes oil, in addition to the gas mentioned in the quote.

When I was approached about a book in this space, I viewed it as an opportunity to draw attention to the deficit in available measurement capability and do something to address it. I was in the enviable position of advising two exciting developments that were right on point. One is a shoebox-size mass spectrometer with the resolution of a laboratory machine, thus enabling detection and speciation of all manner of molecules, not the least of which are volatile organic chemicals (VOCs). VOCs are not currently regulated at well site perimeters. In fact, only recently, in 2016, has there been promulgation of benzene regulation on refinery perimeters. The portable mass spectrometer could well be a case of better measurement emboldening legislation, as well as voluntary compliance.

The second development is the use of subsurface DNA sequencing to characterize oil and gas reservoirs through studies of microbial populations. This is expected to improve recovery economics, thus addressing another leg of the sustainability stool. My co-author, Rob Knight, a world authority on microbiomes in humans and developer of many of the key data analytical techniques, was vital for doing justice to that part of the book.

Fugitive methane emissions from oil and gas operations are being scrutinized, and the issue is somewhat controversial. The Environmental Defense Fund has taken a lead in funding the quantification of the problem. It and the US Advanced Research Projects Agency–Energy (ARPA-E) of the US Department of Energy have funded innovation to improve detection methods. Both the quantification and the developments are described in some detail.

In the face of a future with low oil prices, sustainability importantly includes measures to drive down the cost per barrel produced. Although the price to play certainly includes environmental compliance, resiliency to severe drops in the world price is required. The analytical methods in support of improving recovery economics comprise an interesting blend of chemistry, geochemistry, and biochemistry. Nevertheless, there is no suggestion of a comprehensive recipe for success. This is merely an attempt to fill the toolbox of the folks looking to address this objective.

The entire book has a solutions flavor, in the belief that regulations are more likely to be adhered to if cost-effective alternatives exist. Voluntary compliance will also be encouraged. Process means to monetize the gas that would otherwise be flared is an example of a solution that is not a simple substitution of better performing equipment. In general, the book is intended to be technical but approachable.

Vikram Rao

ACKNOWLEDGMENTS

Many contributed with references, reviews of the text, and other manner of support. Those with significant contributions are identified by chapter here. All comments that follow are those of the lead author (VR).

Jason Amsden (Duke University) provided much insight and material, including figures, for Chapter 4, Particulate Matter and Volatile Organic Chemicals. David Vinson (University of North Carolina at Charlotte) conducted an exhaustive and informative review of early drafts of Chapters 5, Methane in Groundwater, and 6, Potential for Liquid Contamination of Groundwater, keeping me honest on the geochemistry. Joel Walls (Ingrain Inc.) made valuable contributions to the text, and provided many of the figures, for Chapter 7, Illuminating the Reservoir. Luke Ursell and Nicole Scott (Biota Technology) were very influential with regard to Chapter 9, Subsurface DNA Sequencing: A New Tool for Reservoir Characterization. Jonathan Thornburg, Dorota Temple, and Brian Stoner (all RTI International) provided key insights for Chapters 3, Detection of Methane and Amelioration, and 4, Particulate Matter and Volatile Organic Chemicals.

Gerald T. Pollard (Howard Associates, LLC) patiently edited the copy; his lasting contribution was my introduction to Strunk and White, no doubt a feeble attempt at reducing the load of future copy editors. Dayle Johnson (RTI International) endured artistic suggestions from the engineer author and outdid himself with a brilliant cover design.

CHAPTER 1

Background

That which does not kill me, makes me stronger.
(Was mich nicht umbringt, macht mich stärker).

—**Friedrich Nietzsche, *Twilight of the Idols***

Shale oil and gas may have permanently altered the energy landscape. They burst upon the fossil energy scene with a suddenness that initially defied prediction. Much of the uncertainty was due to the fact that the type of reservoir being exploited was dramatically different from the conventional one. The term "unconventional reservoir" was ascribed to this rock. The relatively new techniques of horizontal drilling and hydraulic fracturing were essential. Both of these techniques, but especially the latter, had some environmental baggage, especially in states such as Pennsylvania that were unprepared with adequate regulations. Production methods were inefficient largely because the reservoir was not adequately understood. This book addresses the analytical methods and associated science necessary to permit efficient exploitation while simultaneously protecting the environment. Had methods such as these existed in the early 2010s, and regulations emboldened by them put in place, much of the uncertainty engendered to date could have been avoided.

OIL

A scant 6 years ago, not even the most aggressive crystal ball gazers could have predicted the events of 2014, when shale oil production ramped up rapidly enough to be the determining factor of world oil price. By late 2014, the price of oil halved. Prices in early 2016 plumbed

Sustainable Shale Oil and Gas. DOI: http://dx.doi.org/10.1016/B978-0-12-810389-0.00001-2

new recent depths down to US$25 per barrel. The Organization of the Petroleum Exporting Countries (OPEC)'s response was to maintain production and not prop up the price with production cuts. A widely reported motivation of at least Saudi Arabia, a member of OPEC, was to drive out higher cost shale oil producers. Shale oil production certainly had dropped by early 2016, but resiliency crept in. Production costs were slashed in part with cost reduction by the service companies but mostly through efficiency innovation. An important component of this game of chicken is a realistic appraisal of how low the profitable breakeven cost of shale oil can go. The time frame is a factor as well, as described later.

What, then, are the prospects for shale oil producers surviving the Saudi gambit? We treat the natural gas issue separately. The markets are different, and the only material effect of oil price is on liquefied natural gas (LNG) pricing; this is discussed later. The survival of US shale oil is dependent on three factors. The main factor is the ability to produce it profitably at depressed oil prices. The second is the ability to overcome environmental hurdles, because doing so may represent the price to play. Lastly, shale oil production, unlike conventional oil recovery, literally occurs in the backyards of citizens who are not used to this activity. These three factors add up to the standard definition of sustainability: Without profit, there is no enterprise, but it cannot be at the expense of either the environment or the wellbeing of the citizens in the locality of the industry.

As a consequence of the foregoing, this book squarely addresses all three elements of sustainable production. This hydrocarbon source is new and, consequently, nowhere close to optimized with regard to economics of production. For example, only approximately 5% of the oil in place in the average play is being produced. The percentage for conventional oil, by contrast, is in the mid-thirties. Modest gains in the 5% figure will have a significant impact on profitability. However, these gains require capabilities not currently in the toolbox of practitioners. During the period 2013 to 2015, the industry focused on improving efficiency in logistics and drilling practices. It is now common for multiple wells to be placed on pads. The following box describes this phenomenon, which has benefits to the environment in addition to the economics of drilling.

Multiwell Pads

Fig. 1.1 is a schematic of a typical multi-well pad. The principal feature is the ability to drill multiple wells from a single location. Typically, but not always, the rigs in the pad move on rails to the new location. This takes several hours, compared to several days for normal operations. Conventional practice requires a rig to be partly disassembled, moved, and reassembled. Considering that wells are currently being drilled and completed in less than 20 days, the savings in time are highly material with multi-well pads. An operator in Colorado drilled 50 wells from a single pad on 4.6 acres and drained a reservoir of 640 acres. Environmental benefits include less surface disturbance for the same subsurface acreage accessed and fewer access roads for the transport of water, sand, and other fracturing related materials. A larger pad also has the critical mass to affordably install processes for environmental compliance, including water treatment facilities and air emissions testing equipment. The Colorado example is unusual in its size. Most pads have 4 to 15 wells, in part because the subsurface accessible from a large pad may not be under lease to the operator of the wells.

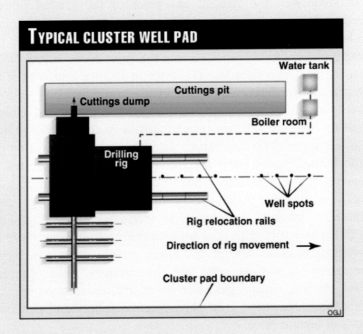

Figure 1.1 A multiwell pad. Courtesy *Oil and Gas Journal.*

NATURAL GAS

Natural gas is a regional commodity, unlike oil, which is fungible and has a world price. The exception is the relatively recent phenomenon of LNG. Chilling natural gas, primarily methane, to $-162°C$ at atmospheric pressure converts it to a liquid that can be transported long distances if kept at approximately that temperature. This is achieved by deliberately boiling off small quantities. The latent heat of evaporation chills the liquid. In ships, the resultant gas is retrieved and used as fuel. At the destination, LNG is returned to the gaseous state in regas terminals. All three steps together have a considerable cost. Natural gas is priced in US$ per million British thermal units (MMBTU). A thousand cubic feet (Mcf) of natural gas has an energy content of approximately 1 MMBTU, so the market price is often quoted as US$ per Mcf.

The delivered LNG price will add between US$4 and US$6 per MMBTU to the production cost, depending on factors such as distance shipped. This effectively ensures a regional element to pricing. Due to shale gas, the natural gas price in the United States has declined to between US$2 and US$4 during the last couple of years, and it has recently been at the lower end of the range. The US Energy Information Administration forecasts the price to be below US$6 for at least the next two decades. The price in Europe and Asia is driven by the price of the "last cubic foot," which would be LNG sourced. As a result, products with a high natural gas component, as either raw material or fuel, are severely disadvantaged in those areas relative to the United States. This has caused an influx of manufacturing capital into the United States from abroad. In the chemical industry alone, new capital of approximately US$150 billion has been committed. In a few years, when these plants are up and running, natural gas demand will increase. In fact, it already has increased: Natural gas is now (early 2016) in parity with coal for electricity production. Despite this displacement of coal, natural gas prices remain depressed. New uses are necessary for a substantial rise. Importantly, unlike the case of shale oil, the low natural gas price is self-inflicted. Abundance caused the imbalance. No cartel is in play. Normal supply and demand economics will dictate price.

The aforementioned applies only to the United States. In countries in which LNG sets the marginal price, the plummet in oil price has had a profound impact. In most countries, medium-term delivery contracts for LNG are pegged to the price of oil. Thus, in 2015, the price dropped in concert with the drop in oil prices, and the future price remains uncertain. The price in Asia was up to US$19 per MMBTU in early 2014; in early 2016 it was close to US$9. All of this still leaves the United States with a considerable advantage in the manufacture of chemicals sourced from natural gas. Consequently, US prices will likely firm over time. The significance is that innovation in cost reduction may not be as critical for gas as it is for oil. Demand creation is more important. The challenges are similar, and the oil side may have done the gas industry a favor. For oil, it is a matter of innovate or perish. Gas production will be a beneficiary. On the environmental side, the issues are similar but possibly slightly more acute for gas because gas is more mobile than oil. The gains to be made using methods in this book and elsewhere will certainly apply.

We have chosen to focus on techniques that enable better illumination of the reservoir. This book is not intended to be a comprehensive treatise on reducing the cost to produce a barrel of oil. Although that is certainly the most appropriate metric, we focus simply on new and emerging analytical techniques that enable that objective. The two techniques featured rely on recent advances in data acquisition and associated data analytics. One uses sophisticated electron microscopy to estimate reservoir rock properties. The other uses the microbiome to identify rock with greater production potential. The second is derived from existing data analytics associated with the human microbiome and on the fact that DNA sequencing costs have plummeted by four orders of magnitude in recent years.

Formation evaluation methods perfected over the last 80 years or so are now seen as applicable primarily to conventional reservoirs. To further confound matters, the high cost of formation fracturing left less room for costly reservoir estimation operations such as sophisticated logging in horizontal wells. The result was the aforementioned low recovery rate. While oil prices were north of US$100 per barrel, there was not much incentive to improve recovery factors. Furthermore, we most certainly do not imply that the methods described herein are the only means for improving reservoir understanding. We hope the book will stimulate further research.

ENVIRONMENTAL CONSIDERATIONS

These fall roughly into two buckets: water and air. Each well can consume up to 6 million gallons of water. The water is mostly fresh, so the industry competes with other users, which is especially pernicious in recent drought conditions. Only about 10 to 35% of injected water returns to the surface as flowback. The water left behind can affect reservoir performance, but it is flowback that has the potential to contaminate freshwater sources and is therefore the basis for much of our discussion. Many of the analytical techniques for assessing contamination are not new, but some are not widely known in the industry. Their applicability is discussed in detail, especially in the geochemical context. For example, in the event of aquifer contamination by methane, one can use these techniques to predict the age of the rock from which the offending fluid was derived. They can also decipher whether the source was terrestrial. This has implications for culpability and remedial action.

Air emissions from shale oil and gas operations require significant investigation. Analytical techniques continue to be inadequate. For example, volatile organic chemicals are not currently measured at the perimeter of operations because doing so is economically not feasible. Existing studies are largely phenomenological with regard to health outcomes in proximal populations. We describe an emerging technology with the potential to identify and quantify airborne organic molecules precisely at a multiplicity of locations at relatively low cost. Also described is the measurement of particulate matter (PM10 and PM2.5), which is implicated in morbidity and mortality. One of the underlying beliefs occasioning this book is that cost-effective analytical techniques could inform legislation and possibly embolden it. In its June 2015 decision, the US Supreme Court gave clear instruction to the US Environmental Protection Agency to take into account the economics of enforcement prior to promulgation of rules.

Fugitive emissions of methane have faced intense scrutiny since 2011. New analytical methods have emerged and more are being researched to identify sources and quantities. Direct release to the atmosphere is increasingly uncommon, in part due to the scrutiny. However, flaring (combustion) is the norm when no profitable outlet can be found for the gas. The highest volumes of flaring are for gas associated with oil production and for which no export pipelines exist.

Inefficient burners result in release of aromatic compounds such as benzene in addition to carbon dioxide. Because harnessing of gas that would otherwise be flared has the potential to create economic value, some discussion of recent advances in that space is included. Viable economic alternatives are powerful incentives for compliance.

Many factors—some predictable, others not—feed into this new phenomenon of shale petroleum with its technological, economic, environmental, geopolitical, and social ramifications. The advances in analytical methods described herein are likely to have implications well beyond the oil patch.

PART I

Air Emissions

And if you gaze for long into an abyss, the abyss gazes also into you.
(Und wenn du lange in einen Abgrund blickst, blickt der Abgrund auch
in dich hinein.)

—*Friedrich Nietzsche*, **Beyond Good and Evil**

CHAPTER 2

Fugitive Methane and Emissions From Flaring

Shale oil and gas production has resulted in two types of atmospheric emissions that have recently come under scrutiny. Fugitive methane is usually defined as gas that is released to the atmosphere inadvertently. We add to that the category of deliberate release, in part because it is a significant component. One mechanism is a result of using equipment that inherently leaks when operated; whether that may be classified as deliberate is partly in the realm of semantics. In many instances, especially in the case of gas produced in association with oil, the gas is flared instead of vented. In either case, anthropogenic greenhouse gas is added to the atmosphere. Some of these practices have always been present even in production from conventional reservoirs. The emissions from flaring of associated gas are still far greater worldwide from non-shale sources than from shale operations. In this chapter, we detail the sources, and in Chapter 3. we discuss the analytical methods required to detect, quantify, and ameliorate them.

Because this field of inquiry is new, we address recent mitigation measures. Another reason to discuss economical mitigation is that sustainable production could well require it in some measure. If the leakage rate exceeds a certain proportion of total production, the net effect of shale gas on the environment could well be negative compared to alternatives.[1] This will surprise many because gas is viewed as cleaner than other hydrocarbons, certainly coal. That is correct: Burning gas produces less than half the carbon dioxide as burning coal. (Of course, coal has other important emissions, such as particulate matter and mercury, which are absent from natural gas. Note, however, that NO_x capture in the combustion process is done with ammonia, and a small amount of nitrates of ammonia can result, which are particulate matter.) We report on the quantification of this difference by the Environmental Defense Fund (EDF), in particular its calculation of the point at which the amount

Sustainable Shale Oil and Gas. DOI: http://dx.doi.org/10.1016/B978-0-12-810389-0.00002-4

of produced gas lost to the atmosphere begins to negate the inherent advantage of gas. To some degree, this limiting state may not be relevant because it is known from an EDF-sponsored study that conventional technology alone, if implemented, can mitigate a very high fraction of the emissions very economically.[2] Innovative technologies are being actively sought, some of which are alluded to in Chapter 3. Collectively, therefore, one may be entitled to some bullishness on shale oil and gas as a net positive for the environment compared to the alternatives.

SOURCES OF METHANE LEAKAGE

Methane may leak out at every stage in the production and delivery system. Fig. 2.1 is a sketch of the system from the point of production to the end use. Not shown, but discussed later, is natural seepage on land and in the ocean, mostly methane hydrates, which are plentiful. According to a US Geological Survey estimate, the total potential gas from this source is greater than that from all fossil fuel production operations combined.

In the simplest terms, the chain begins at the well sites. These may be individual wells or, as is becoming increasingly common, a cluster of wells on what is known as a pad (see Chapter 1). The gas may be from a well designed to produce natural gas or from an oil well that also produces some gas. This last is known as associated gas. The gas is sent away from the rig site on gathering lines. These are small-diameter pipelines, usually well under 10 inches in diameter, as opposed to interstate pipelines, which are 36 or 40 inches in diameter. Although small, gathering lines have an installed cost of more than US$1.5 million per inch-mile. Thus, each mile of a 5-inch line costs approximately US$7.5 million. This high cost is the reason some lines are deferred until later in the life of the site when there is more certainty of the economic viability of the prospect.

Gathering lines usually connect to a processing plant. Here, the larger liquid molecules are removed as natural gas liquids, usually by chilling. Propane, butane, and natural gasoline are separated and sold. Some ethane is also removed. Pipeline specifications limit the thermal content of gas to 1.1 million British thermal units (MMBTU) per thousand cubic feet (Mcf) of natural gas. Ethane

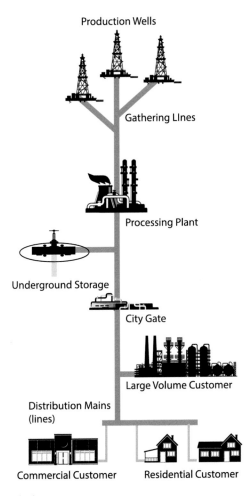

Production Wells

Gathering LInes

Processing Plant

Underground Storage

City Gate

Large Volume Customer

Distribution Mains
(lines)

Commercial Customer   Residential Customer

Figure 2.1 Natural gas supply chain.

typically has an energy content of 1.78 MMBTU per Mcf compared to a nominal 1.0 MMBTU for methane, so even a small percentage of ethane raises the calorific value. Since early 2014, ethane has not had a ready market, with prices sometimes lower than that for natural gas. This is in part because, unlike propane and butane, it has no consumer use. The only utility is as a raw material for the manufacture of ethylene, and ethylene capacity in the United States is low compared to the ethane supply. Furthermore, all but 3 of the 39 ethane crackers are on the Gulf Coast, far from much of the ethane production. Producers will put as much ethane as allowable into the pipeline.

Upon exit from the processing plant, the gas goes to a compression station to allow long-distance transport. There are four principal destinations, as shown in Fig. 2.1. Factories using natural gas as a fuel, such as steel mills, and those that use it as a raw material, such as methanol- or ammonia-producing plants, get their supply directly from the interstate line. Power plants also get it directly. A third destination is storage facilities, usually underground, commonly topped up before the winter months. Finally, there is the "city gate," from which there are multiple destinations, including industry, homes, and local storage. In some cases, power plants are also fed from beyond the city gate.

This detailed description of the supply chain is given because most of the elements are associated with fugitive emissions, but with different mechanisms, and, more important for this book, often with different methods of detection and quantitation.

PRODUCING WELLS

The well pad has two classes of methane leakage: equipment-related and as a consequence of the fluids-extraction process. In the first category are hatches and covers of fluid-filled containers with inadequate sealing. Valves and gaskets are also classified in this category. The fracturing process involves injection of high-pressure fluid to fracture the rock and induce artificial permeability in the reservoir. At some point, fluid returns to the surface, and this is characterized as flowback. Flowback water will bring with it a certain quantity of natural gas from the reservoir rock. The mixture is sent to a separator tank where the gas is easily removed. In many instances, it is either used on site for power generation after some conditioning treatment or sent to the gathering line for export to the separation plant. Early in development of the site, gathering lines may not have been put in place. In this case, the operator has two choices: release to the atmosphere or burn in a flare. The more environmentally benign option of flaring is increasingly being chosen. Quantification of this is ongoing in the EDF-sponsored study at the University of Texas at Austin.[3]

The second category of emissions is the operation known as liquids unloading. Occasionally, the production from a gas well is impeded by the accumulation of associated liquids at the bottom of the well. A common means of removing them is by blasting the liquid with high-pressure gas from the reservoir. When the fluids mixture gets to the top, the gas portion is vented. An alternative that avoids much of this fugitive emission is the use of the plunger lift. This method calls for shutting in the gas production and then inserting the plunger to the bottom. When the well is opened, the seals around the plunger allow pressure to build below it, and this lifts it. The accumulated liquid rises and is collected. All the energy for the lift comes from the reservoir pressure, so the method is economical. Aside from the environmental benefit, little gas is lost from sale, although there is some fugitive emission with current designs. A 2012 survey showed that less than half of wells use this method, so there is room for better practices. "Room for better practices" is a constant refrain.

Pneumatic devices in the flow system use the gas pressure in the line to actuate devices such as valves. Some of these require a small amount of the gas to bleed out as a design feature. Low-bleed devices are classified as emitting less than 6 standard cubic feet per hour, and high-bleed devices are classified as emitting greater than 6 standard cubic feet per hour, often as much as 30. Some devices, such as fluid level controllers, are intermittent. The usual remedy for this is to use only low-bleed ones. Actuation by compressed air instead of the high-pressure gas in the line is best environmentally but is more expensive.

TRANSMISSION LINES

The transmission system commences at the gas processing plant, which can have some fugitive emissions. However, the largest contributor is the next step in the line, the compressor station. The gas is compressed to make it flow the distance required. The fuel for compression is almost always gas from the pipeline, and the combustion produces CO_2 and some unburned methane. However, most emissions are from leaks in various parts of the system, especially the compressor seals, and during maintenance. As a class, this is the worst actor in the chain.[2] Many alternative designs for mitigation have been put forth. They generally are more expensive, so policy action may be required to minimize this source. The next worst actor is pneumatic actuation devices, but, as discussed previously, means of mitigation exist at relatively low cost.

One of the destinations along the transmission line is storage. Most common is storage in a field that produced gas at one time but is now depleted. This is the lowest cost means. The fields are usually distributed close to transmission lines and hence are convenient. The most versatile are salt caverns. These are usually near the Gulf Coast. Salt bodies can be hollowed out by circulating water through them. Salt caverns are also used for oil storage in the Strategic Petroleum Reserve. Neither of these storage means is particularly cited as an avenue for fugitive emissions other than from pneumatic valves operating on gas pressure.

The other destination for gas is power plants. Sometimes they are on the other side of the city gate.

BEYOND THE CITY GATE

Industrial and domestic customers are supplied. Urban distribution systems are strongly implicated in fugitive emissions, especially in older emplacements that use unprotected steel and cast iron piping that has deteriorated. These account for 50% of all emissions and 70% of emissions in the eastern United States. A recent study[4] showed that city methane releases are dramatically lower than those listed in the most recent 2011 US Environmental Protection Agency (EPA) report because that report relied on a comprehensive study done in 1992. Fig. 2.2 shows the updated values.

The findings of Lamb et al.[4] are not surprising for the equipment sector because one would expect improvements, especially when in the direct control of industrial companies as opposed to consumers. The watchword is leaked gas is gas that is not sold. The findings also imply that significant replacement of old pipes has occurred. Lamb et al. also note that eastern US cities are in much worse condition. Note the very large error bars in Fig. 2.2, representing high uncertainty in the statistics. This is very likely due to the small samples compared to the EPA numbers, which are voluminous but old. Recent studies have emerged for Washington, DC, Boston, and New York that paint a bleaker picture.[5,6] However, where pipes have been replaced, the improvements have been significant, as demonstrated in three cities in the eastern United States.[7]

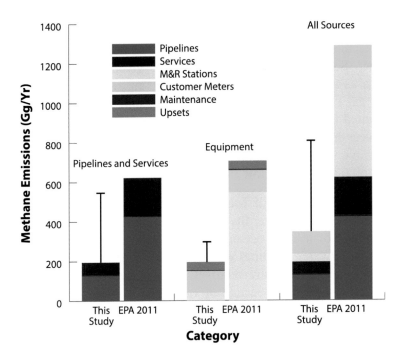

Figure 2.2 Updated US inventory of methane emissions from local distribution systems. Lamb BK, Edburg SL, Ferrara TW, Howard T, Harrison MR, Kolb CE, et al. Direct measurements show decreasing methane emissions from natural gas local distribution systems in the United States. Environ Sci Technol 2015;49 (8):5161−5169

METHANE HYDRATES AND THEIR ROLE IN GLOBAL WARMING

Methane hydrates are ice-like molecules containing methane. They are ubiquitous and plentiful. According to the US Geological Survey, they are by far the most plentiful hydrocarbon source in the world. They are discussed in this book because they are known to seep at rates that have not been properly quantified. In part because many of the deposits are in marine formations, the seepage is not easily detectable. There is a belief in many circles that global warming will accelerate the seepage. The possible mechanism for this will become evident later.

Methane hydrates are also known as methane clathrates. These are very large molecules, not unlike the buckyball form of C60, whose discovery by Smalley, Curl, and Kroto was recognized with the Nobel Prize in 1996 and is the basis for major advances with carbon nanotubes. The lattice comprising water (ice) has interstices filled with methane molecules.

A significant portion of the volume of hydrate comprises the methane molecules. As a result, 1 L of hydrate has approximately 169 L of methane. Some consideration has been given to using methane hydrates as a mode of gas transportation, an alternative to liquefied natural gas (LNG). British Gas and Aker Engineering, among others, investigated the idea, but no commercial system exists. Aker published a complete design of a system and demonstrated that hydrate transport was feasible up to 3500 nautical miles at ordinary freezer temperatures for a capital cost 25% below that of LNG.[8]

Methane hydrates will form when the temperature and pressure allow (Fig. 2.3). In nature, they are found almost exclusively in permafrost on land or in deep ocean sediments. In commerce, they routinely form in underwater pipelines when a certain pressure is exceeded. This is one of the reasons why relatively small pockets of gas (less than approximately 50 MM barrels of oil equivalent) are not developed; they are too small for a platform and cannot be transported to a hub very far because of hydrate formation. A very significant portion of gas resources in the US Gulf of Mexico is stranded for this reason. The first "hat" that BP put on the Deep Water Horizon blowout to contain the oil spill was clogged with hydrates.[9]

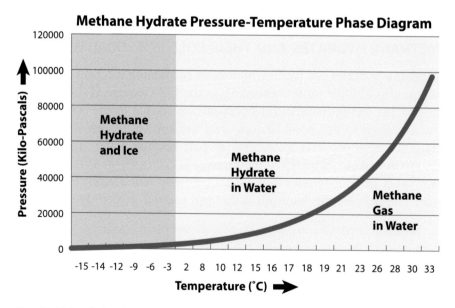

Figure 2.3 Methane hydrate formation conditions. Source Wikipedia.

The permafrost deposits are being identified by some scientists[10] as vulnerable to overall earth warming. At any given location, an increase in temperature could dissociate the molecule. In fact, this mechanism is used to keep gas pipelines from clogging—simply keeping them warm. The other common method is to inject methanol or ethylene glycol. Each liter of hydrates will dissociate to produce approximately 169 L of free gas. The theory goes that global warming will cause massive releases of methane from the relatively shallow deposits in permafrost. The evidence to date does not support this hypothesis.[10]

The hydrate deposits in the deep ocean are the most prevalent, accounting for more than 90% of the total. Seeps are unlikely to be affected by earth warming. Also, the molecules are not likely to make it to the surface of the ocean without being oxidized by resident bacteria. This produces CO_2 and adds to that burden in the ocean, which acts as an important sink for atmospheric CO_2. This process is also not expected to be a material factor.

FLARED GAS

We classify flared gas as gas that is combusted at a production location because it is logistically or economically stranded. A small variant is methane from landfills. As explained in Chapter 3, such methane is known as biogenic and is formed by the action of methanogenic bacteria on organic matter. Similar bacteria are active in the rumen of farm animals and are responsible for that significant source of anthropogenic methane. The essential absence of ethane or larger molecules is an effective marker for the source being biogenic. The following box explains the origins of hydrocarbon molecules associated with oil and gas production.

Why Gas Is Found in Association with Oil

All oil and gas were formed from organisms, mostly of marine origin, that were converted by the action of pressure and temperature. The temperature had to exceed 60°C for the action to begin, and the first product was a material known as kerogen. The pressure was induced by the deposition of sediments, usually carried by flowing water. All of this occurred in the Carboniferous Period and slightly before. The earliest hydrocarbon source is in rock dated to the Ordovician, more than 440 million years old. Most of the reservoirs in the unconventional space are younger, dating to the Silurian and Devonian, the latter ending approximately 360 million years ago. This departure into the seeming esoterica of geology has a purpose. One reason to pay attention to the age of source rocks is that geochemists have identified carbon isotopic and other markers that indicate the origins of the fluids. This recently came into play when methane intrusions into aquifers were reported[11] as being from Marcellus drilling activity. The provenance was disputed by others[12] based on analyses such as these. This is discussed in Chapter 3.

Kerogen is acted on by pressure and temperature to form a hydrocarbon. The formula for this is C_nH_{2n+2}, where n can be from 1 to more than 100. If $n = 1$, we have methane; if $n = 2$, ethane; and so on. If n is greater than approximately 25, we call the mixture of these molecules oil. Oil is the first useful fluid to form out of kerogen. Actually, in some places that lack conventional oil, such as Estonia, kerogen near the surface is mined and processed with heat to yield a useful oil. (As a point of interest, Estonian oil shale is from the Ordovician.) However, this process is energy intensive and believed to be environmentally suspect.[13] With greater thermal maturity, the oil continues to break down into ever smaller molecules. Shale oil is light, representing a late stage in the maturation of kerogen. The lightness of an oil is best characterized by what happens when it is boiled. The lighter the oil, the greater the proportion of useful products from simple distillation. The residue in this case is known as fuel oil and can be sold as boiler fuel.

Light oil undergoes further conversion to ever lighter molecules and finally to methane, which is the most thermally mature state of the original kerogen. Because this is a continuous process, at any given stage, one is likely to have a mixture of fluids. This is the crux of the reason why most oil, but light oil in particular, will have gases in association. We discuss heavy oil later and explain why it is unlikely to have associated gas. Similarly, a natural gas reservoir will have in association some larger molecules. These are usually ethane, propane, and butane, with ethane (C_2H_6) being in the highest proportion, which makes logical sense once one understands the mechanism of formation: It is the second most thermally mature fluid. Gas with a high associated liquid content is known as wet gas. By contrast, if there is little or no liquid, it is dry gas; the outstanding example is the prolific Haynesville field in Louisiana.

Heavy oil is a special case of bacterial action modifying the fluid constituents. Species of anaerobic bacteria feed on hydrocarbons. They prefer the lighter molecules. The net result is that, over time, the oil comprises a higher concentration of larger molecules. One would expect, therefore, that heavy oil will have little or no associated gas. The following is an interesting observation: Oil seeps in the ocean result primarily in tar balls showing up on beaches. Ocean bacteria have consumed the smaller molecules preferentially. Tar is at the heaviest end of oil.

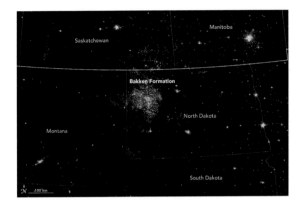

Figure 2.4 Nighttime visibility of North Dakota gas flares. Courtesy NASA.

Associated gas is disposed of in one of three ways. The obvious out-let is sale. In places such as Alaska, where natural gas has no commer-cial outlet, it is reinjected into the oil reservoir, where it serves the useful purpose of increasing the pressure on the oil because the gas is resident in a "gas cap" above the oil layer and presses down on the oil, increasing the production rate. If this is not feasible, the gas is either released (increasingly uncommon) or flared. In relatively remote areas such as North Dakota, a high fraction is flared. Fig. 2.4 is a nighttime image of the United States. Easily observable is the flaring in the Bakken fields, appearing as bright as a major city.

In late 2014, North Dakota flared approximately one-third of asso-ciated gas produced, amounting to 375 MM standard cubic feet per day (SCFD), which even at the depressed gas prices of early 2016 had a value of approximately US$1 million a day. The economic loss is sig-nificant, and simply building more gas pipelines is not the solution. In part, this is because natural gas prices are expected to stay depressed with an oversupply. In fact, dry gas prospects are simply not being put on stream. Furthermore, if oil prices remain depressed as expected, oil production will drop, likely idling some of the new gas export lines. Thus, new gas lines may not pay off. In addition, the gas is widely dis-tributed, making aggregation and dispatch more costly. Any solution to monetize the gas in some way other than shipping to users must rec-ognize this feature. Fig. 2.5 illustrates the situation. Note that the vast majority of well pads are flaring less than 100,000 SCFD. For the measures one could take to ameliorate the situation, this is a signifi-cant stumbling block because processes for natural gas conversion to

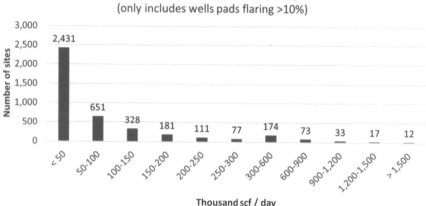

Figure 2.5 The wide distribution of gas flaring in North Dakota. Courtesy RTI International.

useful chemicals are generally run in large plants requiring more than 100 times that quantity. The high volumes are needed to take advantage of economies of scale. Chapter 3 discusses new developments aimed at turning this problem into an opportunity.

QUANTIFYING FUGITIVE METHANE

Howarth et al.[14] estimated the fugitive methane to comprise 3.6 to 7.9% of all the natural gas produced. As discussed in relation to the work of Alvarez et al.[1] quantification is important because natural gas as a substitute for coal or transport fuels loses its advantage if the leakage is greater than a certain proportion of production. Howarth et al. identified the same sources in the value chain that we identified previously in this chapter. The EDF commissioned a series of studies to quantify emissions from the various parts of the chain. The ongoing study led by Allen at the University of Texas at Austin is the most definitive to date on emissions at the production site. They made direct measurements at 190 onshore sites in the United States, comprising 489 wells. Although this is a small set compared to the EPA national inventory of methane emissions, the investigation is more current and more comprehensive where conducted. It highlighted the shortcomings of current analytical techniques. This resulted in EDF sourcing of innovative ideas and the subsequent funding of new methods, as described in Chapter 3. It also encouraged the US Advanced Research

Projects Agency–Energy's MONITOR (Methane Observation Networks with Innovative Technology to Obtain Reductions) program, which is also described in Chapter 3.

The results from Allen et al.'s study[3] are striking in that the completions phase, including flowback water handling, is relatively benign. That estimate is more than an order of magnitude less than the EPA figure and substantially less than the estimate of Howarth et al.[14] This could be explained in part by the fact that with all the scrutiny, better practices are now being followed, certainly when watched. Cynicism aside, the good news is that it can be done with no real economic impact. If all the fluid is collected, separation into gas and liquid phases is simple. Then it is only a matter of how the gas is handled thereafter. The fact that methane release can be kept so low should embolden legislation, as discussed in Chapter 10. The only remaining hurdle for such legislation is cost-effective monitoring.

The data for the other gas handling methods at the rigs are not as promising. Pneumatic controllers reported by Allen et al.[3] had nearly double the EPA values despite the absence of high-bleed valves in the testing group. Equipment leaks in devices such as pumps are also major contributors. More important, there is high regional variability. The positive aspect of that statistic is that current practice could be modified to be uniformly less polluting, much as has been achieved in flowback water handling. Allen et al. did not estimate the effects of liquids unloading, and in computing the total emissions from upstream operations, they assumed the data from the EPA national inventory of 2012. A later study by many of the same investigators confirmed the EPA data to within a few percentage points.[15] Fig. 2.6 details the relative contributions from the various sources. The contributions from equipment design, including those associated with liquids unloading, are striking, and they highlight areas of emphasis for mitigation.

Allen et al.[3] reported the net emissions as 0.42% of the gross natural gas produced. The EPA figure is 0.47%. For the same category of emitting sites, Howarth et al.[14] estimated between 2.2% and 4.0%; this involved very few actual measurements and relied mostly on data from the literature. This metric can be important in deciding the role of natural gas in displacing incumbent fuels in industry. Alvarez et al.[1] tackled this task; their findings are described next.

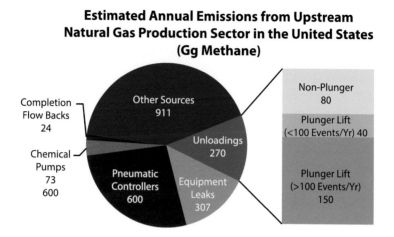

Figure 2.6 Sources of methane emissions in the upstream and their relative contributions. Courtesy American Chemical Society.

CLIMATE CHANGE IMPACT OF NATURAL GAS SUBSTITUTION OF FUELS

The climate impact of each candidate fuel must be calculated on a "wells-to-wheels" basis, meaning cradle to grave, from production to final consumption. For coal, this is "mine-to-combustion." The key illustrations for our purpose are the plots of net climate benefit achieved by natural gas substitution as a function of percentage fugitive emissions (Fig. 2.7). This allows better estimates of the metric as well as improvements to and incorporation of technology. A higher slope is preferred because the benefit of natural gas over the alternative can be realized in fewer years. For example, at a leakage rate of 3%, the gasoline substitution will see the benefit in slightly more than 40 years. The same point for diesel substitution would be 140 years. The coal comparisons are made between combined cycle natural gas and supercritical combustion coal plants using low-CH_4 coal, essentially the best case scenarios for both. The striking conclusion is that the transport fuel substitution currently most in practice is the most sensitive to fugitive emissions. Of course, this is the substitution of compressed natural gas for diesel. There is a caveat: This substitution also essentially eliminates particulates, both PM10 and PM2.5, which are health hazards. In a similar vein, Alvarez et al.[1] discuss the fact that coal combustion often leads to sulfur-based emissions, which have a net cooling effect on the environment. They discard this from their

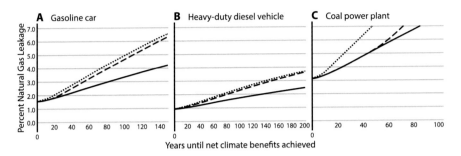

Figure 2.7 Maximum fugitive methane emissions for fuel substitution with climate benefits. Single emissions pulses are represented by dotted lines. The dashed lines represent the service life of the vehicle or plant, and the solid lines are for permanent fleet substitution. Courtesy Alvarez RA, Pacala SW, Winebrake JJ, Chameides WL, Hamburg SP. Greater focus needed on methane leakage from natural gas infrastructure. Proc Natl Acad Sci 2012;109(17):6435–6440.

analysis because they believe that the health outcomes are great enough to force regulators to limit sulfur even in countries that do not currently limit them.

Substituting natural gas for coal appears to be the best scenario in this analysis. At slightly more than a 3% loss, natural gas remains advantaged over coal, not even considering the collateral health damage of coal production and use. To put in perspective the viability of achieving any of the three minimum levels in Fig. 2.7, one must first understand the broad industry segments involved. The value chain can be divided into upstream, midstream, and downstream. In Fig. 2.1, all of the elements from the rig site through the gathering lines are classified as upstream. Elements from the gas processing plant to the city gate are midstream. Everything after is downstream, which also includes the power plants before the city gate. The significance of the divisions is that each tends to have a different type of owner. Upstream is oil and gas companies and supporting service companies, which supply most of the equipment and services. Midstream is a different set of companies skilled either in pipelines or in chemical separation of fluids. Downstream is utilities retailing to consumers or chemical plants producing petrochemicals.

The reason for keying on the companies in each segment is that each has different competencies to appreciate the methane leakage imperative and, more important, different core competencies in being able to take ameliorative action. The other point is that midstream and downstream are in principle the same regardless of whether the gas comes from shale or conventional reservoirs. In fact, one could go further and argue that even intrapad gas handling is common to conventional gas rigs as well. The distinctions are in the processes and procedures uniquely associated with hydraulic fracturing. According to the Allen study, these have seen the most gains in changing practices, and the worst actors are the pneumatic controllers and equipment leaks, which are common to all sources of gas production. The study computes the total upstream emissions today to be 0.42% of the total natural gas produced, leaving a fair amount of space from the minimums in Fig. 2.7, especially if fixes are viable for the worst equipment actors.[2] When the leaks in the midstream are taken into account, this figure will be higher. This observation brings another fact into focus: The midstream exists regardless of the source of the natural gas. To the extent that this book is about sustainable production of shale oil and gas, the upstream numbers are more germane.

Other countries attempting to produce shale oil and gas would do well to note the key takeaway from the previous discussion: Existing infrastructure is suspect on fugitive emissions. This applies whether or not they intend to produce unconventional hydrocarbons. This aspect is discussed in Chapter 10. The United States would likely not have been aware of these facts had it not been for shale oil and gas industry scrutiny.

To estimate what could be feasible for amelioration, the EDF commissioned a study by ICF International.[2] It updated the EPA 2011 baseline numbers with recent data. It then performed a detailed examination of every unit operation in the value chain and estimated the feasibility and cost of amelioration in each case. Any derived economic benefits were separated into those for the operator and those for the public at large. Reasonably, only the operator was viewed as directly applicable to decisions to mitigate methane emissions. Helping mitigation is the fact that the 80/20 rule is applicable: 22 of the more than 100 emission sources are responsible for greater than 80% of the emissions projected for 2018. Importantly, the report concluded that onshore methane emissions could be reduced by 40% with current technologies and methods for a net cost of US$0.66 per Mcf of methane emission reduced and less than US$0.01 per Mcf of gas actually produced. This takes into account the economic value of the gas prevented from leaking accruing to the implementer of the improvements.

Although the actual changes are likely to be executed by a service company, the cost and benefit can reasonably be ascribed to the oil and gas operator or a midstream entity.

There is an element of irony in all of the foregoing. Hydraulic fracturing drew the attention of people such as Howarth as a likely bad actor in methane emissions. However, the ensuing scrutiny revealed that the fugitive methane ascribable to fracturing is a fraction of that in the rest of the value chain. Most of this chain has been around for decades with conventional production, and we now know that it has problems. The idea that one would have pneumatic control systems that leak gas by design might come as a surprise to the public. The irony is that but for the antifracturing sentiments, the rest may well have not been found out.

There are more than 100 points in the gas value chain where methane release is possible. Current methods for detection fall short in identifying and quantifying the sources. Efforts to fill this gap are discussed in Chapter 3.

REFERENCES

1. Alvarez RA, Pacala SW, Winebrake JJ, Chameides WL, Hamburg SP. Greater focus needed on methane leakage from natural gas infrastructure. *Proc Natl Acad Sci* 2012;**109**(17):6435–40.

2. ICF International. *Economic Analysis of Methane Reduction Opportunities in the US Onshore Oil and Gas Industries*. <https://www.edf.org/sites/default/files/methane_cost_curve_report.pdf>; 2014 [accessed 07.07.16].

3. Allen DT, Torres VM, Thomas J, Sullivan DW, Harrison M, Hendler A, et al. Measurements of methane emissions at natural gas production sites in the United States. *Proc Natl Acad Sci* 2013;**110**(44):17768–73.

4. Lamb BK, Edburg SL, Ferrara TW, Howard T, Harrison MR, Kolb CE, et al. Direct measurements show decreasing methane emissions from natural gas local distribution systems in the United States. *Environ Sci Technol* 2015;**49**(8):5161–9.

5. McKain K, Down A, Raciti SM, Budney J, Hutyra LR, Floerchinger C, et al. Methane emissions from natural gas infrastructure and use in the urban region of Boston, Massachusetts. *Proc Natl Acad Sci* 2015;**112**(7):1941–6.

6. Jackson RB, Down A, Phillips NG, Ackley RC, Cook CW, Plata DL, et al. Natural gas pipeline leaks across Washington, D.C. *Environ Sci Technol* 2014;**48**(3):2051–8.

7. Gallagher ME, Down A, Ackley RC, Zhao K, Phillips N, Jackson RB. Natural gas pipeline replacement programs reduce methane leaks and improve consumer safety. *Environ Sci Technol Lett* 2015;**2**(10):286–91.

8. Gudmundsson JS, Børrehaug A. Frozen hydrate for transport of natural gas. Second International Conference on Natural Gas Hydrate, June 2–6, 1996, Toulouse, France. <http://www.ipt.ntnu.no/~ngh/library/paper3.html>; 1996 [accessed 29.02.16].

9. Borowsky L. BP oil spill prompts hydrate research. *Mines.* 2010 Fall/Winter. <http://mine-smagazine.com/388/>; [accessed 14.02.16].

10. Ruppel CD. Methane hydrates and contemporary climate change. *Nat Educ Knowl* 2011;**3**(10):29.

11. Osborn SG, Vengosh A, Warner NR, Jackson RB. Methane contamination of drinking water accompanying gas-well drilling and hydraulic fracturing. *Proc Natl Acad Sci* 2011;**108**:8172−6.

12. Molofsky L, Connor J, Farhat S, Wylie Jr A, Wagner T. Methane in Pennsylvania water wells unrelated to Marcellus shale fracturing. *Oil Gas J* 2011;**109**(19):54−67.

13. Mittal AK. Unconventional *Oil and Gas Production. Opportunities and Challenges of Oil Shale Development* [GAO-12-740T]. US Government Accountability Office. <http://www.gao.gov/assets/600/590761.pdf>; 2012 [accessed 14.02.16].

14. Howarth RW, Santoro R, Ingraffea A. Methane and the greenhouse-gas footprint of natural gas from shale formations: a letter. *Clim Change* 2011;**106**:679.

15. Allen DT, Sullivan DW, Zavala-Araiza D, Pacsi AP, Harrison M, Keen K, et al. Methane emissions from process equipment at natural gas production sites in the United States: liquid unloadings. *Environ Sci Technol* 2015;**49**(1):641−8.

CHAPTER 3

Detection of Methane and Amelioration

In Chapter 2, we reported the current state of understanding of the sources and amounts of fugitive emissions of methane. Also noted were the statistics on natural gas flaring. The vast majority of natural gas is flared when there is no commercial outlet for sale. Flaring is classified into two categories: offshore conventional associated gas and onshore gas associated with shale oil. The potential means for amelioration may be different because the offshore volume is more than an order of magnitude greater than the onshore volume.

The most appropriate method for detection of methane is dictated by the specifics of the source: at the production and processing sites or during pipeline transport. In the case of the former, the likely zones of leakage are generally known, and it is a matter of devising sensors and a protocol. For pipelines, the demands are greater because of the distances and interfering factors. One such factor is the possibility of biogenic methane from farm animals. Accordingly, some sort of speciation is required to detect the presence of the larger molecules, which essentially could only come from thermogenic sources. An assist comes from the fact that pipeline gas is allowed a quantity of ethane limited only by the total calorific content of 1150 British thermal units (BTUs) per thousand cubic feet (Mcf). Accordingly, it will almost always have molecules larger than methane, differentiating it from biogenic gas from landfills, cattle, and so forth.

ARPA-E AND EDF PROGRAMS FOR DETECTION

The US Advanced Research Projects Agency–Energy (ARPA-E) unit of the US Department of Energy and the Environmental Defense Fund (EDF) each launched programs to develop sensing systems to address the fugitive methane problem. EDF had already been responsible for spearheading the elucidation of the problem, as discussed in Chapter 2. In both programs, the emphasis is not so much on cutting-

Sustainable Shale Oil and Gas. DOI: http://dx.doi.org/10.1016/B978-0-12-810389-0.00003-6

edge detection methods as it is on the balance of detection and quantification with systems cost. The focus is on systems cost, not simply
capital cost. Systems cost is defined as the fully loaded cost of operating the system for 1 year. This is in part because in some cases the service may be provided by a third party, with the equipment not
necessarily owned and operated by the oil and gas company.

The ARPA-E program is called Methane Observation Networks
with Innovative Technology to Obtain Reductions (MONITOR). The
goal is a detection threshold of 1 ton per year, which corresponds to
1.9 grams per minute or approximately 6 standard cubic feet (SCF) per
hour. This is defined for a well pad of 10×10 meters, with leaks permitted at any location on the pad. In recognition of the possibility that
some methods would employ distributed sensing with a means to compute the leak source, they specified an average wind velocity of
2.75 meters per second. Even with point sensing, wind effects may be
modeled to minimize the number of locations of testing. The overall
goal is to reduce methane release by 90% for an annual fully loaded
cost of less than US$3000 for what they refer to as basic functionality,
which is defined as methane detection without further speciation of
larger molecules or organic compounds. Additional functionality is
permitted a higher cost per year. Further requirements are for identifying the source of the methane with 1-meter resolution and communication of the results wirelessly to a specified location.

Seven of the 11 projects now funded are described here as examples
of the strategies followed. All but one are based on methane absorption spectra in the infrared range. Fig. 3.1 demonstrates the principle.

All of the developments described here use methane absorption
spectra, usually in the near infrared (0.7–1.4 μm), short-wave infrared
(1.4–3.0 μm), and mid-wavelength infrared (3.0–8.0 μm). The basic
science underlying the measurement is roughly the same, and the innovations involve size, mobility, and cost. A key feature is the choice
between mobile and stationary. If stationary, then reduced cost is
usually the objective. Stationary units are augmented with systems to
combine multiple measurements to locate the leak source. In all cases,
the objective is fitness for purpose and conformance to the price/
performance goals set out by the MONITOR program. This last is
typical of ARPA-E funding priority; the basic science is generally left
to other divisions in the agency. Mission orientation is basic, as it is

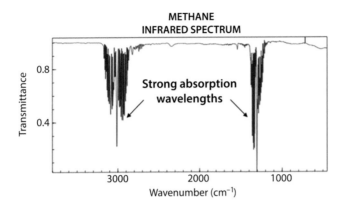

Figure 3.1 Two strong absorption peaks of methane in the infrared spectrum that are used as detectors of the gas. The figure uses the units of wave number. The conversion is: wavelength [μm] = 10000/wavenumber [cm⁻¹] NIST Chemistry WebBook (http://webbook.nist.gov/chemistry). Ironically, this light-absorbing characteristic of methane is the reason why it functions as a greenhouse gas in the atmosphere.

for the US Department of Defense's Defense Advanced Research Projects Agency program, on which ARPA-E is modeled.

Improved and/or Lower Cost Components of the System

- Lead partner General Electric has a hollow-core optical fiber for long path-length transmission of infrared radiation at methane absorption wavelengths. Micrometer-sized holes in the fiber admit gases into the hollow core, allowing location and quantification of the methane. A distinguishing feature is that the method is applicable not only at the well pad but also in gathering lines and pipelines (described in Chapter 2).
- The team led by Maxion Technologies targets cost reduction of laser systems. Its objective is a 40-fold reduction. The instrument will be a tunable laser, with wider application, which should yield economy of scale. If tunability allows detection of ethane as well as methane, then speciation for biogenic/thermogenic differentiation should be possible. In the case of methane, the team is aiming for resonance frequencies in the vicinity of 3.3 μm.

Mobile Systems

- The Bridger Photonics system is to be mounted on either a ground vehicle or an unmanned aerial vehicle. It is intended to survey a 10-meter-square pad in 5 minutes. The novelty is in the use of the near-infrared fiber laser amplifier and the data analytics associated with the global positioning system and inertial navigation to yield three-dimensional imagery of the methane absorption. This localizes the leak points and allows a focus on problem spots. The system would serve multiple well pads to reduce the overall average system cost.

- Physical Sciences leads a team to redesign a laser-based detector to fit in an unmanned aerial vehicle. It claims the detector will measure ethane as well, so at least two sets of resonance frequencies are likely to be investigated. The novelty is that the flight is automated based on the initial survey; the system has designed intelligence to then change the flight path and zero in on the suspected leak with higher precision. The team intends to survey the perimeter of the oil and gas site. That has limited utility for methane (because perimeter measurement is not called for, the value being in detecting the source and reducing the leakage), but the functionality should allow the higher value survey of gathering lines.

- Rebellion Photonics has the most aggressively novel mobile technology. The crux is a methane plume imager communicating with cloud computing. A lightweight camera is deployed on the personal protective equipment of a worker, probably the hard hat. Each pixel of the image is populated with multiple bands of spectral data. From the standpoint of deployment, this is the most interesting of the MONITOR technologies. Routine inspection and maintenance schedules for normal rig operations can be repurposed for methane detection with minimal effort. Consequently, methane monitoring would be minimally intrusive to rig operations. This attribute in particular would appeal to oil and gas operators.

Other Methods

- The IBM-led project attacks the problem at two levels. The first is to produce very low-cost sensors based on tunable diode laser spectroscopy. The principle is to focus on a single resonance frequency and tune the diode to that one frequency. The second is to miniaturize into an all-on-one-chip version, which is expected to use less power and be one or two orders of magnitude lower in cost. These sensors will be distributed over the pad and connected by a self-organizing mesh. The idea is the polar opposite of the Rebellion mobile method.

- The Duke University and RTI International method is a complete departure from all the others in the program. A shoebox-size mass spectrometer has been devised to have the full resolution of a room-size machine. The conventional slit aperture is replaced with a coded aperture comprising multiple slits of varying widths and spacing. Visually, the effect is that of a diffraction grating, and in fact the coded aperture serves a similar function as the grating in an optical application. Being a mass spectrometer, it allows a high degree of speciation, identifying and quantifying all larger molecules, including ethane. This ability to deal with large molecules makes it useful for identifying volatile organic compounds (VOCs). Accordingly, it informs much of Chapter 4 and is not further discussed here. Despite the elegance and power of the technique, it is unlikely to compete favorably with some of the others mentioned here for methane alone and would be most similar in functionality to the Bridger method except that it is hand carried rather than worn. If the mission is combined with VOC control, this might be considered more favorably. Currently, however, there are no regulations for VOCs at well sites, and even the fugitive methane regulations are still new (see Chapter 10).

GREEN COMPLETIONS

The US Environmental Protection Agency (EPA) promulgated new regulations covering VOCs and methane in 2012.[1] The effective date for full compliance was set as January 2015 to allow retrofit and refurbishment. Central to this change was the requirement for green completions at all wells for the primary purpose of producing natural gas. The rules did not, at the time, apply to gas associated with oil production. A green completion, formally known as reduced emission completion, is one in which, following the completion operation including fracturing, the flowback natural gas is captured and not vented. The captured gas may be cleaned up for sale or used for another purpose, such as the production of local power. Until the 2015 date, flaring would be permitted, but not direct release.

Even in 2012, more than half of all wells did not intentionally vent any gas. However, some did flare the gas. More to the point, the equipment for capturing and storing already existed. Dehydrators and three-phase separation equipment are the key pieces required to accomplish the separation. If this equipment is present and properly sized, the total cost for compliance is US$700 per day of use. If it has to be installed, the cost is nearly an order of magnitude more.

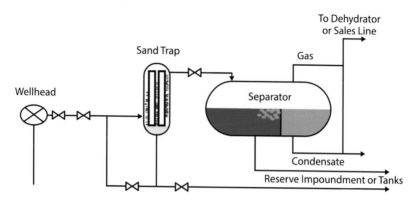

Figure 3.2 Green completions schematic. Courtesy: Green Completions IPIECA (http://www.ipieca.org/energyefficiency/solutions/78161).

Importantly, the equipment is only needed for the few days of flowback prior to actual production. This duration varies but is a matter of days. Consequently, portable equipment supplied by a third party would be advantageous unless the overall volume over time justifies purchase. Fig. 3.2 is a schematic of the process. The sand trap removes fine solids, and the three-phase separator removes condensate and water separately. The condensate is sold, and the water is treated for use. The gas is sent to a dehydrator, and dry gas is then disposed of.

Such portable equipment is on offer by industry, so compliance will be strictly a cost issue. If a use can be found for the natural gas, such as pipeline export or power production, a net benefit is achieved, according to the EPA.[2] Later, we describe a US Department of Energy-funded technology in late-stage development that converts natural gas into a liquid fuel on the small scale dictated by this application.

The regulations also concern many of the sources of leakage identified in Chapter 2. For control valves using line pressure, low bleed valves with a designed leak rate of 6 SCF per hour are a requirement. For certain other valves, a zero bleed rate is specified. As discussed in Chapter 2, the solutions are all commercially available. The regulations ignore an important leak source—the liquids unloading process described in Chapter 2.[3] As noted in that chapter, the alternative plunger lift process is available and cost-effective. Accordingly, this may be a good example of allowing voluntary compliance while still requiring overall mitigation quantitatively.

EPA regulations target 90% reduction in emissions. The residue after the processing shown in Fig. 3.2 is water and volatile organics such as benzene. The latter are simply combusted. In fact, any gaseous residue is combusted. This raises the overall emission reduction to 95%. Since mid-2015, there has been some criticism (and effective rebuttal) of the methane estimation techniques as not being properly quantitative.[4,5] If mitigation is achieved at these levels, it may not much matter whether certain measuring devices were flawed. However, clear identification of the sources of leakage remains important. The MONITOR program and EDF-sponsored investigative activity seek to address this issue.

ALTERNATIVES TO FLARING GAS

Because natural gas has commercial value, it is flared only when it cannot be monetized. In the case of shale gas wells, such a situation could arise in the early days of the prospect development. Before the location is known to be an economic producer, often export pipelines are not laid down. After fracturing the formation, some natural gas returns with the flowback water. It is usually separated from the water in much the same equipment as shown in Fig. 3.2. However, if no gathering lines are ready to take the gas away, it may be vented or flared. Venting is rare now. Consequently, whether to flare is a timing issue confined to the period before gathering lines are constructed.

Gas associated with oil production involves a somewhat different issue. In this case, gas export lines will not exist per se because the fluid being produced for sale is oil. Gas lines may be purpose built if the economics dictate. Fully 30% of North Dakota-associated gas is flared because the economic case cannot be made. Fig. 2.6 highlights the problem: The majority of the wells are producing less than 50,000 SCF per day. At mid-2016 prices, this would sell for just US$100.

Small-scale compressed natural gas (CNG) and liquefied natural gas (LNG) are on offer. In the case of CNG, the units are on trucks that drive to the location, remove the gas, and take it to a pipeline terminal. This "virtual pipeline" also must contend with the delivered price of the gas. Small-scale LNG generally does not extend down to 50,000 SCF per day. The World Bank commissioned a study to examine alternatives to flaring that involved conversion to liquid fuels. It published a report in 2013 and updated it in 2014. The further update in 2015[6] moves some of the companies around according to their near and long term potentials, but the message remains roughly the same, especially regarding near-term availability.

The World Bank study is widely quoted in the industry. The term MiniGTL designates any technology converting natural gas into a liquid fuel or chemical of value. This is a broader definition than in the past, when the liquid could only be diesel or gasoline. In all cases, the feed stock is methane, sometimes mixed with larger molecules. In most cases, the targeted feed rate is 1–20 MM SCF per day (SCFD). These rates apply to many situations, including offshore oil platforms with no gas offtakes. For the much smaller rates needed by onshore-associated gas flaring, the World Bank cites several companies whose technology is in early development. Greyrock, with a more near-term capability, has an offering at 0.5 MM SCFD targeting the flaring opportunity,[7] producing what is stated to be wax-free diesel.

The product liquids are usually methanol, dimethyl ether (DME), gasoline, and diesel. The last two are conventional transport fuels and consequently very saleable. Methanol and DME have chemical markets and only recently have appeared as gasoline and diesel substitutes, respectively, especially in China. They depend on the following chemical reactions:

$$\text{Syngas production: } CH_4 + H_2O \rightarrow CO + 3H_2 \quad \Delta H = +206\,kJ/mol \quad (3.1)$$

$$\text{Methanol from syngas: } 2H_2 + CO \rightarrow CH_3OH \quad \Delta H = -92\,kJ/mol \quad (3.2)$$

$$\text{DME from methanol: } 2CH_3OH \rightarrow CH_3OCH_3 + H_2O \quad \Delta H = -23\,kJ/mol \tag{3.3}$$

$$\text{Shift reaction to increase } H_2/CO \text{ ratio: } CO + H_2O \rightarrow CO_2 + H_2$$
$$\Delta H = -41\,kJ/mol \tag{3.4}$$

$$\text{Fischer–Tropsch for olefins: } 2nH_2 + nCO \rightarrow C_nH_{2n} + nH_2O$$
$$\Delta H = -165 - 180\,kJ/mol \tag{3.5}$$

All of the processes begin with a mixture of CO and H_2, which is known as synthesis gas (syngas). Methanol from syngas is exothermic, and heat removal is important for control of the reaction. In comparison, the Fischer–Tropsch (F-T) process is highly exothermic, as the numbers show on the enthalpies of reaction in Eq. (3.5). The desired product for our application in mitigating flaring is olefins. However, due to the difficulty of controlling the exotherm, quantities of wax (longer chain molecules) are produced, which has to be cracked to a transport fuel or sold separately. This added complication in equipment and the business model makes any process for small-scale production in remote areas challenging. However, at least one of the companies on the World Bank inventory claims to produce wax-free diesel. In principle, this would qualify as a candidate for remote locations.

DME production from methanol is a very simple dehydration process, as shown in Eq. (3.3), and has a mild exotherm that can be controlled easily. However, a hydrocarbon-tainted water waste is produced, with 0.5 mol for every 1 mol of methanol, which is nearly 30% by weight. In a remote location, this too is contraindicated. The methanol-to-gasoline process that converts DME to gasoline catalytically is also a complication that removes the process from consideration.

As a practical matter, methanol is the most appropriate liquid target (other than wax-free F-T). It is an easily transportable commodity with a world price and multiple end uses. Furthermore, methanol synthesis is less sensitive to the $H_2:CO$ ratio than F-T. In the latter case, the ratio must be close to 2, and if the syngas synthesis falls short, then the shift reaction described in Eq. (3.4) must be employed, which is again a complication relative to equipment in the field.

For processes starting with syngas, all the innovation has been on the back end, with better catalysts and the like. There are other methods for syngas production beyond the steam reforming shown in Eq. (3.1), such as dry reforming using CO_2 as a reagent. However, all the alternatives will have an even higher relative capital cost, especially on a small scale. One report noted that reforming equipment comprised 60% of the capital cost of a DME production plant in China.[8] Innovation targeted at reforming appears merited.

SYNGAS PRODUCTION USING A DIESEL ENGINE

A diesel engine is an internal combustion engine in which a carbonaceous fuel is combusted to produce energy to power a drivetrain.

$$CH_4 + 2O_2 \rightarrow CO_2 + 2H_2O \tag{3.6}$$

Eq. (3.6) shows the use of methane rather than diesel fuel. When methane is used, spark ignition is required, rather than the normal compression ignition. In an automotive engine, full combustion of the fuel is sought, so the engine is run lean—i.e., with oxygen in some excess over that required by stoichiometry. Because combustion air contains nitrogen, this has the negative effect of producing oxides of nitrogen (NO_x), but that is tolerated if a trap is used to capture the species. One such trap is to react the gas with injected urea in the tail-pipe, yielding nitrogen as the product. Most trucks use this technique and replace the urea canister at gas stations. The stoichiometric reaction is

$$4NO + 2(NH_2)2CO + O_2 \rightarrow 4N_2 + 4H_2O + 2CO_2 \qquad (3.7)$$

The following box describes an alternative approach, which was the basis for controversy involving the automobile maker Volkswagen.

The Volkswagen NO$_x$ Episode

NO_x is a "front-of-the-box" pollutant. The effects are short term and on health, in contrast to CO_2, the effects of which are long term and on targets such as severe weather and drought, leaving the public with some room for doubt about causality. Consequently, NO_x emissions from devices such as automobiles could be expected to be a public concern. Yet much of the attention from the Volkswagen (VW) emissions episode is directed to the company's behavior and not the attendant pollution. In fact, the reporting has shown that much of the industry has indulged in avoidance in one way or another. In Europe, emissions testing is done by the manufacturer with no regulatory oversight.[9] The use of nonstandard vehicles during testing is a common practice to which most people turn a blind eye. For mileage testing, cars are routinely stripped of wind drag components such as side view mirrors. Real-world miles per gallon can be 25 to 50% worse than in these tests.[10]

When a diesel engine is tuned to burn lean, it performs best, especially with respect to fuel economy. This condition is defined as air slightly in excess of the stoichiometric amount required to combust the fuel. Less unburnt fuel also means minimal release of unburnt hydrocarbons. However, the excess air causes increased production of NO_x. This must be reduced in the exhaust gas stream.

VW uses a technology known as the Lean NO_x Trap (LNT). There are two steps (and a preliminary step that we skip for simplicity). In the first, NO_x is captured on a coating that *adsorbs* NO_x. Adsorption is a surface phenomenon that is easily reversed. When the coating is considered filled up, the second step kicks in, which involves removing the NO_x to regenerate the coating activity. This is the key step that got VW into trouble. The NO_x is reduced to nitrogen and CO_2 on a special catalyst by reacting it with a mixture of hydrocarbons, hydrogen, and CO. This mixture is created by switching the engine to a rich burn mode, away from the lean. The reactant is the fuel from the cylinders that is only partly combusted (note the similarity of this mechanism to the method described later in the chapter: the deliberate use of a diesel engine to produce syngas). Not surprisingly, during this time, engine performance declines because it is being operated in suboptimal conditions. Fuel mileage declines, as does torque.

It appears that VW decided to defeat the device during normal road operation. If this was in fact done (we are not here to stand in judgment; this is a technical discussion of what could be), then a reasonably sophisticated algorithm would be needed to detect that the vehicle was in a test mode. When in this mode, the engine would be allowed to run rich for the needed period to perform the function of the LNT. Importantly, however, in normal driving the vehicle would run lean all the time, giving the needed performance in miles per gallon and torque. In other words, it was peppy (high torque) with high mileage and was great on measured emissions. Keep in mind that all diesel engines are better on mileage than gasoline engines. This is in part because the fuel has approximately 10% more energy content and in part because diesel engines run at much higher compression ratios. For decades, however, they have had a reputation for being smoky and smelly. That is no longer the case. Particulate filters take away the smoky aspect. The only remaining concern is the NO_x emissions.

Much as in the case of the LNT, the diesel engine methane reformer runs rich, defined as oxygen deficient. The stoichiometric combustion of CH_4 in air has a fuel:air ratio of 17.2, derived from the fact that methane has 16.04 g/mol and air has 28.96 g/mol. The richness is defined by the equivalence ratio φ_A:φ_S, where S is stoichiometric, and A is for actual. This is denoted as φ_{HC} in the figures. At stoichiometric conditions, the ratio is 1; in lean conditions, it is <1; and in rich conditions, it is >1. When rich mixtures are run, the reaction products are primarily CO, H_2, CO_2, and N_2. The desired quantity is syngas, the first two components.

The following discussion regards development work at the Massachusetts Institute of Technology,[11] with subsequent scale-up and further testing by RTI International and Columbia University, all funded by ARPA-E.[12] The principal objective is to target flared gas volumes. Based on the North Dakota data shown in Fig. 2.5, the target flow of methane has to be as low as 50,000 SCFD. For other applications, such as biogas from landfills and animal waste impoundments, targets of 300,000 SCFD appear merited. Accordingly, the project is designing for an 8-liter diesel engine for the lower volumes and two 13- to 15-liter engines for the higher volumes. Both embodiments, including the methanol conversion, are planned to fit on two flatbed trucks. The portability would also allow short-duration applications in the early stages of development of shale gas wells. The primary objectives of the development are as follows:

- H_2:CO ratio of 1.6 to 2.0 (The lower figure is adequate for subsequent conversion to methanol, and the latter is required for F-T synthesis.)
- First-pass conversion of methane greater than 50%
- Simulation of anticipated recycling of H_2 adding up to 5% in the inlet gas
- Evaluation of the ability to handle up to 20% ethane mixed with the methane
- Determination of the optimum equivalence ratios for each of these conditions
- Syngas cleanup process to allow input to a methanol or F-T reactor
- Production of methanol from the reformer output (A third party is designing the small-scale methanol conversion unit.)

Many of these objectives are dependent variables. The ethane requirement is there for two reasons. First, most pipeline gas has some ethane, often up to 4%. Producers use ethane to make up to the required 1 MM BTU per Mcf of natural gas and are permitted to take it to the pipeline limit of 1.1 MM per Mcf. Second, gas associated with shale oil has significant proportions of natural gas liquids. The mechanism of formation of oil and gas would predict this outcome for light oil.[13] In the Bakken, the natural gas liquid component is in the range of 8 to 12 gallons per Mcf of gas.[14] Of this, approximately half will be ethane, approximately 25% propane, and the balance butane and larger molecules. Service companies offer skid-mounted equipment to drop out propane and larger molecules by chilling. However, ethane cannot be separated on location and will need to be handled by any flaring alternative. This is the reason why this or any other solution needs to account for ethane in the feed.

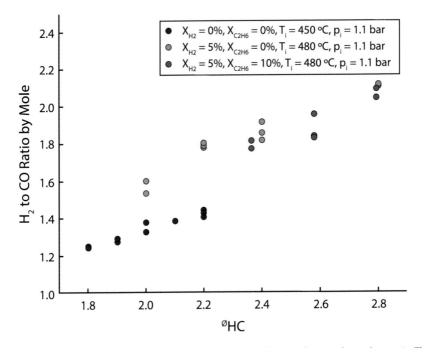

Figure 3.3 Conversion percentages of methane (and ethane when mixed in) as a function of equivalence ratio. The points marked "x" indicate the ethane component alone. Courtesy Massachusetts Institute of Technology.

As expected, the leaner mixtures delivered a lower H_2:CO ratio, with both ethane and hydrogen making it richer and hotter and delivering a higher H_2:CO ratio. An ethane-rich mixture can cause soot formation in the exhaust. Accordingly, an equivalence ratio of 2.2 will likely be the target for delivery to a methanol reactor.

Fig. 3.3 data were from the same runs that produced the data in Fig. 3.4. The conversion efficiency of the fuel mix dropped below approximately 2.2 on the equivalence ratio. The provisional hypothesis is that the hydrogen is being reacted preferentially to the methane. Note that the ethane conversion in isolation remains high, although it, too, is tailing down at the highest ratios. Because H_2 recycling is expected and ethane is likely to be present, a practical limit on the equivalence ratio on the basis of conversion efficiency is 2.2. This is also the figure derived based on the H_2:CO ratio. If this process is used as a precursor to F-T synthesis, which requires an H_2:CO ratio of 2.0, then either the process conditions will be adjusted or the output gas will be subjected to the water–gas shift reaction in Eq. (3.4) to produce the additional hydrogen required.

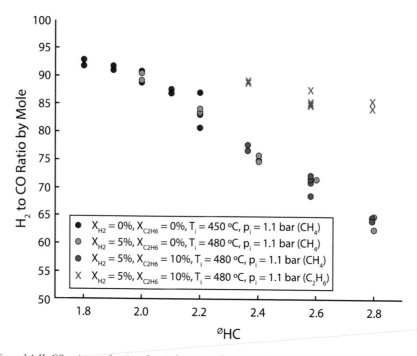

Figure 3.4 *H_2:CO ratio as a function of equivalence ratio for various feedstocks including 5% recycled hydrogen and 10% ethane. Ethane is directly substituted for methane on a volume basis with the total moles of fuel remaining constant. Ethane has almost twice the calorific value of methane.* Courtesy Massachusetts Institute of Technology.

Although ethane substitution data are shown only up to 10%, successful operation was possible up to 20%. Such a high concentration is unlikely in most associated gas. However, ethane is in oversupply in the United States and often sells for less than natural gas despite having nearly twice the calorific content. Consequently, this process could usefully monetize a cheap feedstock while these conditions prevail.

The engine reformer is a good example of distributed chemical processing. It is also an example of economies of mass production replacing economies of scale.[13,15] The concept also benefits from remote monitoring control and from the modularity allowing parts to be short-lived and replaceable. This is in contrast to chemical plants, which are built to last decades.

REFERENCES

1. US Environmental Protection Agency. *Oil and Natural Gas Sector: New Source Performance Standards and National Emission Standards for Hazardous Air Pollutants Reviews.* <https://www3.epa.gov/airquality/oilandgas/pdfs/20120417finalrule.pdf>; 2012 [accessed 19.06.16].

2. US Environmental Protection Agency. *Oil and Natural Gas Sector Hydraulically Fractured Oil Well Completions and Associated Gas during Ongoing Production.* <http://www.slideshare.net/devkambhampati/oil-gas-drilling-methane-vocs>; 2014 [accessed 19.06.16].

3. Allen DT, Torres VM, Thomas J, et al. Measurements of methane emissions at natural gas production sites in the United States. *Proc Natl Acad Sci* 2013;**110**(44):17768−73.

4. Brownstein M. The UT methane studies, critique and response. <http://blogs.edf.org/energyexchange/2015/03/03/the-ut-methane-studies-critique-and-response/>; 2015 [accessed 19.06.16].

5. Brownstein M, Hamburg S. Keeping an important methane research question in proper perspective. <http://blogs.edf.org/energyexchange/2016/06/09/keeping-an-important-methane-research-question-in-proper-perspective/>; 2016 [accessed 19.06.16].

6. Fleisch TH. *Associated Gas Monetization via miniGTL: Conversion of Flared Gas into Liquid Fuels and Chemicals, Report III* [World Bank]. <https://openknowledge.worldbank.org/handle/10986/23609License: CC BY3.0IGO>; 2015 [accessed 19.06.16].

7. PR Newswire. Perseus and Greyrock announce joint venture to monetize flare gas in Mexico. <http://www.prnewswire.com/news-releases/perseus-and-greyrock-announce-joint-venture-to-monetize-flare-gas-in-mexico-300276676.html>; 2016 [accessed 21.06.16].

8. Fleisch TH, Basu A, Sills RA. Introduction and advancement of a new clean global fuel: The status of DME developments in China and beyond. *J Nat Gas Sci Eng* 2012;**9**:94−107.

9. Hruska J. More manufacturers found to violate diesel emissions standards—but blame the test, not the vehicles. <http://www.extremetech.com/extreme/216039-more-manufacturers-found-to-violate-diesel-emissions-standards-but-blame-the-test-not-the-vehicles>; 2015 [accessed 20.06.16].

10. Vidal J. Car manufacturers manipulating fuel efficiency tests, says report. <https://www.theguardian.com/environment/2013/mar/14/car-manufacturers-manipulating-fuel-efficiency-tests>; 2013 [accessed 20.06.16].

11. Lim EG. <http://hdl.handle.net/1721.1/100109> The engine reformer: syngas production in engines using spark-ignition and metallic foam catalysts. *MS thesis.* Mechanical Engineering, Massachusetts Institute of Technology; 2015 [accessed 20.06.16].

12. ARPA-E. Compact, inexpensive micro-reformers for distributed GTL. http://arpa-e.energy. gov/?q = slick-sheet-project/compact-inexpensive-reformers-natural-gas>;   2012   [accessed 20.06.16].

13. Rao V. *Shale oil and gas: the promise and the peril.* 2nd ed. Research Triangle Park, NC: RTI Press; 20152015 [chapters 1 and 18].

14. Salmon R, Logan A. *Flaring up: north dakota natural gas flaring more than doubles in two years* [Ceres report]. <http://www.ceres.org/resources/reports/flaring-up-north-dakota-natural-gas-flaring-more-than-doubles-in-two-years>; 2013 [accessed 20.06.16].

15. Dahlgren E, Lackner KS, Gocmen C, van Ryzen G. DRO-2012-03 Small modular infrastructure. <http://www8.gsb.columbia.edu/rtfiles/dro/VanRyzinSummer2012.pdf>;   2012 [accessed 21.06.16].

CHAPTER 4

Particulate Matter and Volatile Organic Chemicals

Particulate matter (PM) and volatile organic chemicals (VOCs) are the principal classes of airborne pollutants associated with shale oil and gas production. Of the two, VOCs attract the most attention from the public. Much of the reported concern centers on health outcomes believed to be linked to VOC emissions rather than on exposure measurements. This is at least in part because citizens are unable to obtain exposure data for their areas of residence or work. For both classes of pollutants, conventional measurement methods are inadequate. Air monitoring stations, although designed with parameters such as prevailing wind direction in mind, are sparsely distributed and expensive to install and operate. They cannot inform citizens of conditions in the immediate environment.[1] A recent survey by the US Environmental Protection Agency (EPA) concluded that the state of the art was not adequate for cost-effective measurement of critical species in discrete communities.[1] This conclusion applied to both classes of pollutants.

With a different objective, the intelligence community is seeking a means of chemical detection with high resolution and portability.[2] The characteristics of such a device suitable for the petroleum industry are discussed in the section on VOCs.

PARTICULATE MATTER

PM is classified into two categories: PM directly related to materials in and around rigs and PM in the emissions from the use of diesel fuel. The former is primarily cement dust and fine particles from the handling and storage of sand or other proppant. This type of PM is unlikely to have any impact off the site, and onsite exposure is covered by US Occupational Safety and Health Administration rules.

Diesel PM emissions are from either stationary or mobile sources. Almost all fracturing pumps operate on diesel fuel. Their emissions

Sustainable Shale Oil and Gas. DOI: http://dx.doi.org/10.1016/B978-0-12-810389-0.00004-8

occur intermittently, during the fracturing portion of well operations, which is typically a few days. Fracturing is followed by approximately the same number of days of rig movement and drilling operations. Then fracturing will recommence. This is in normal operations. In batch drilling, several wells are drilled but not completed, known in the industry as DUC (pronounced duck) wells. This offers the ability to defer the expensive fracturing step. In mid-2016, it was estimated that several thousand DUC wells were on hold waiting for better pricing before being fractured. In such a design, the emissions from the fracturing pumps are concentrated in time. Wind direction and speed determine the spread to the local area. The stationary (but distributed) PM monitors in the area are unlikely to be useful in identifying local impact.

The mobile source is diesel trucks transporting materials to support the operation. Where local surface water is not available, fresh water is brought in tankers. If the wastewater disposal sites are located elsewhere, as they probably are in the case of disposal in Underground Injection Control Class II wells, wastewater is hauled away in tankers. Thus, mobile sources will vary depending on whether local water is needed or flowback water is used. However, transport of sand or other proppant applies to all operations, and variation lies only in the amount used per stage. There is a trend toward using more, in the belief that it increases recovery.

Truck traffic is episodic. Long convoys produce bursts of emissions over short periods. Information on the frequency and volume of traffic could be acquired by the local municipality from the operator, but existing monitoring sites are unlikely to cover the route taken or the drilling site. Because the objective is to estimate the impact on the local population, the adequacy of PM measurements in other settings could be instructive. Except for a few studies, some of which are ongoing, most reporting has been about retrospective or anecdotal incidences of health outcomes to individuals. This parallels the situation for secondhand tobacco smoke and urban automotive discharge into the air.

The body of literature on PM measurement for evaluation of health outcomes appears unanimous on one point: Results of studies of concentrations measured closely proximal to individuals differ substantially from those of studies of concentrations measured at fixed points, with the latter being the current practice. Only the most evenly distributed contaminants provide similar results from both methods. Sensors close to the breathing zone are the gold standard.[3] This is expensive even if appropriate wearable sensors were available (discussed later). Accordingly, researchers emphasize the need for complementary fixed location testing for cost-effective measures of cohort exposure over long periods.

The shale oil and gas setting needs epidemiologic studies that allow clear lines to be drawn between cause and effect. Cost-effective measurements with fit for purpose accuracy and resolution are needed. Investigators have shown that in these situations, examination of the most exposed portion of the cohort is required.[4] This design is not only sounder but also would define the relative risk between exposed and unexposed populations. In our case, mere identification of the most exposed portion of the cohort may prove challenging because of the mobile source and episodic nature of the releases.

Prospective hydrocarbon areas where little or no activity has commenced offer the opportunity for baseline testing of pollutants. Examples are in North Carolina, New York, and countries such as the United Kingdom, Brazil, and Argentina. Even if the potential exposure areas are identified accurately, which in most cases is likely, the locations of monitors will require a high level of sophistication because of the uncertainties in the source, as noted previously. One such study is underway in North Carolina, near the city of Sanford. The likely production area is well defined, and most of the "bluff bodies" housing the sensors are set up downwind from that area, with one proximal to an existing state-directed air emissions measurement site to facilitate calibration. An effort was also made to estimate the likely routing of trucks carrying materials for the drilling site. Bluff bodies were placed at road intersections and in a school on the route. The sensing is for VOCs and PM. For the latter, the instrument is a MicroPEM, which is a newly developed portable device described later in this chapter. It is one of a group of devices developed in response to the generally held belief that cost-effective measures were not readily available.[5] The National Institute of Environmental Health Services launched an initiative in 2006 to meet this need.[6] A recent review details the state of the art.[7]

SMALL LOW-COST PARTICULATE MATTER MEASURING DEVICES

Personal exposure monitoring is most effective if the system is worn on the body in or near the breathing zone. For this to be accomplished without avoidance behavior, such as removing and laying the device down in a car, the device must be very small. In most instances, there is a trade-off between size and accuracy. Smaller particles tend to be more evenly distributed in the air. Accordingly, for estimating PM2.5, the most dangerous size, the sampler will not need to be very close to the nose. This latitude may even apply to PM10 but not larger. In 2013 and 2014, the EPA conducted an evaluation of all inexpensive (<US$2500) small measuring devices available worldwide at the time of the study.[1] The EPA compared each device against the federal equivalent method (FEM) in short bursts and over a 24-hour period. A Grimm Technologies (Douglasville, GA) model EDM180 PM2.5 (EQPM-0311-195) dust monitor and an R. M. Young model 41382VC relative humidity and temperature sensor comprised the FEM standard. Regression analysis was run on the data from each instrument evaluated. Features such as portability, ease of use, temperature and humidity sensitivity, and time required per measurement were also considered. Table 4.1 shows the results. Only three of the particle counters reached a reasonable threshold of accuracy, based on the r^2 results. Of these, two separated themselves for accuracy and measurement output (micrograms per cubic meter); the third only counted particles. The two are the Met One 831 and the MicroPEM from RTI International. The MicroPEM is singled out in this book for the following reasons: It weighs two-thirds less, it has two to six times longer battery life depending on duty cycle, it is capable of communicating results wirelessly (Bluetooth), it has an onboard accelerometer, and it has the ability to collect particles for gravimetric analysis.

The MicroPEM weighs just 240 grams and is designed to be slung on a camera strap, as shown in Fig. 4.1. The filter is selected to capture either PM2.5 or PM10, depending on the nature of the study, targeting deep lung or thoracic deposition zones, respectively. The particles may be retrieved for speciation and gravimetric analyses. As noted previously, avoidance behavior can compromise the study statistics, so the device has an onboard three-axis accelerometer to validate wearing compliance. Most studies require a minimum 60% compliance, and up to 80% is common. Although the expected application is as a wearable, the device may also be used in stationary mode. State regulators should require that the truck convoy path be disclosed to the regulatory authority a certain number of days prior to the event so that the devices can be deployed quickly. Also, populations along the route can

Table 4.1 Summary of PM Sensor Performance

Sensor	r^2	Response	RH Limit (%)	Temperature Effects	Time Resolution	Uptime	Ease of Installation	Ease of Operation	Mobility
AirBase CanarIT (µg/m³)	0.004	−0.101	100	None	20 s	Excellent	Good	Excellent	Very good
CairClip PM (µg/m³)	0.064	−0.229	95	0.657	1 min	Excellent	Good	Very good	Excellent
Carnegie Mellon Speck (particle counts)	0	0.06	90	None	1 s	Very good	Good	Fair	Good
Dylos DC1100 (particle counts)	0.548	21368	95	None	1 min	Very good	Good	Good	Poor
Met One 831 (µg/m³)	0.773	0.049	90	None	1 min	Excellent	Good	Good	Good
RTI MicroPEM (µg/m³)	0.720	1.35 ± 0.12	95	0.588	10 s	Very good	Good	Fair	Fair
Sensaris Eco PM (µg/m³)	0.315	0.034	100	0.313	Unknown	Bad	Poor	Bad	Poor
Shinyei PMS-SYS-1 (µg/m³)	0.152	0.292	95	None	1 s	Good	Fair	Good	Fair

Source: Courtesy EPA.

Figure 4.1 MicroPEM slung on a camera strap. Courtesy: RTI International.

be narrowed down, and the most exposed portion of the cohort can be identified for possible personal monitoring. If the routes will be permanently retained, possibly as an element in the drilling permit application, then permanent stations can be maintained.

Biomass Cook Stoves Need Improvement

Cook stoves that burn biomass, commonly wood and sometimes cow dung mixed with straw, are used primarily in low- and middle-income countries. Poor room ventilation is common. This results in significant morbidity and more than 4 million premature deaths per year, according to the World Health Organization[8] and the World Bank,[9] almost all in the aforementioned countries. Much has been published on the health effects of PM, especially the smaller PM2.5. Acute lower respiratory infections (including pneumonia), asthma, and cardiovascular disease are demonstrably implicated by several studies. Negative perinatal outcomes such as low birth weight are also reported, as are lung and bladder cancer.[10] The greatest at-risk segment of the population comprises mothers and children. Most studies have used surrogate metrics derived from independent measures of releases from combustion of the implicated fuels. The study reported here, by contrast, obtained its data directly from the home environment. It compares levels of PM2.5 measured in the breathing zone of the cook with levels at fixed monitors in the living space surrounding the cooking operation. A further objective was to compare the use of chimneys and a newer low-emissions Anagi cook stove with the conventional operation. Following exhortations by the World Bank and others, innovations have been forthcoming, including better chimneys as well as stoves with minimal smoke (classed as smokeless). With these improvements have come better study tools for quantifying benefits.

The measurement device in this study[11] was the MicroPEM, worn on a camera lanyard by the cook (as shown in Fig. 4.1) or deployed in the living space. Although it weighs only 240 grams, it may have to be made smaller and lighter for deployment on children. Here, only cooks wore it. Results from the personal units and the fixed units did not differ significantly, likely because of the smallness of the space. However, the chimney produced a marked improvement over the conventional stoves. The Anagi stove was better, but not as strikingly so as the use of chimneys. To inform future studies, two major points were made:

- The remarkably high mean participant wearing compliance of 86% indicates the acceptability of the weight and form factor of the MicroPEM, at least for this population of cooks. Most had some years of high school. Some village settings in other countries will have lower levels of education.
- The overall valid data capture rate of 97% confirms the robustness of the device in this near equatorial environment with relatively high particle loading. Also validated were the adequacy of the training and the ability of the device to be deployed with minimal supervision.

VOLATILE ORGANIC COMPOUNDS

In the oil and gas industry, the VOCs of greatest interest are in the class of aromatics known as BTEX: benzene, toluene, ethylbenzene, and xylenes. The likely locations of emissions are the drilling and production sites, the trucks hauling materials, and facilities such as refineries and chemical plants. Of these, only the last two are covered by sector-specific EPA regulations. Until recently, refinery rules were for operations only inside the fence. In 2016, rules were promulgated for fence-line monitoring of benzene.[12] Because almost all refineries have roads on the perimeter, a portable device would be suitable for routine measurement. Modeling could predict the best quadrant for measurement on any given day based on wind direction and speed. Portability would facilitate traversing just these areas. Fixed monitoring would have redundant measurements. Inside the refinery or chemical plant, portability is an advantage as well. Odor or other indicators can be followed up with immediate measurement. Odor is also often used by the surrounding community as an indicator of undesirable emissions (aromatics, almost by definition, have an odor; the name derives from the word "aroma"). A portable device would aid investigation by both refinery operators and citizen scientists.

Chemical plants, such as ethane crackers and ammonia producers, have parallel issues and equal applicability for a portable identifier and quantifier of organic molecules. Forensics with a portable device has been identified by the intelligence community as important for the detection of explosives and the like. Accordingly, in 2016, the Intelligence Advanced Research Projects Agency (IARPA) issued a request for proposals in this space. Their opinion of the state of the art is shown in an adapted version in Fig. 4.2.

One conclusion to be drawn from Fig. 4.2 is that some type of mass spectrometer is the best candidate for modification to bring it into the lower right quadrant. An old but still relevant review describes the different types of mass spectrometers and their principles of operation, features, and limitations.[13] Another conclusion is that the highest performing devices are large—at least desktop size, if not room size.[14] The IARPA goal is to be handheld and yet perform at the upper end of the forensic scale. This is consistent with our application except possibly for the definition of handheld. A forensic investigator of cargo may well prefer a true handheld capability. In a refinery, certainly

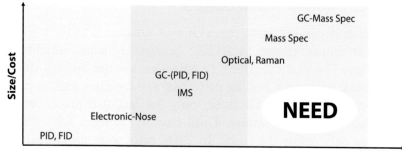

Figure 4.2 *State of the art in measurement of airborne chemical species. The third segment to the right is in the forensic range. The MAEGLIN target is shown by the "need" oval. The CAMMS-ES instrument described in this chapter is expected to fall just above the middle of the oval.*

perimeter measurements would tolerate a somewhat larger object. Portability, rather than handheld capability, may be the defining requirement. A touchstone on weight might be the maximum allowable in a suitcase for stowage on an aircraft, which is 40 pounds. A device would need to balance size, weight, performance, and cost. That balance could be different for differing applications in a single industrial sector. For example, a refinery detector would need to be more precise but could cost more, possibly US$60,000. A detector for an oil well pad or perimeter might have to cost less than US$40,000 but may not be need to be as versatile. These prices are significantly lower than those of laboratory instruments with similar capability.

PORTABLE HIGH-PERFORMANCE MASS SPECTROMETER

This discussion centers on sector mass spectrometers. Sector instruments are the geometry of choice in laboratory isotope ratio mass spectrometers because of superior sensitivity and precision.[15] Note that the performance is in the acceptable forensic range of Fig. 4.2, but the size has to be reduced. The hurdle is to reduce the size while not losing performance characteristics. Miniaturization often forces the instrument designer to confront a trade-off between instrument throughput (and associated signal-to-background ratio) and spectral resolution.

In a conventional instrument, the material being examined is ionized, and the resulting ions are accelerated and leave the ion source through a slit. They enter an area where they are subject to a magnetic field (Fig. 4.3). This causes the ions to separate by mass-to-charge

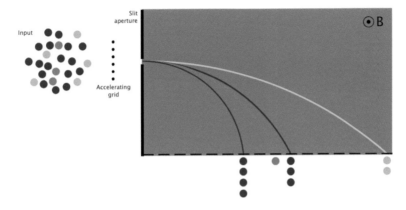

Figure 4.3 Conventional mass spectrometer with single slit. Ions separate by mass-to-charge ratio and go to discrete spots on the detector plate. Courtesy: Duke University.

ratio, which is the characteristic that determines their position on the detector plane. Throughput and resolution depend on the slit size. Therefore, when miniaturizing, one must either sacrifice throughput to maintain resolution or sacrifice resolution to maintain throughput.

Borrowing concepts from optics, a solution to the throughput versus resolution trade-off has been devised for sector mass spectrometers by using spatially coded apertures. Instead of a single slit after the ion source, spatially coded apertures use an array of slits (Fig. 4.4). The position at which the ions hit the detector plane now depends jointly on the mass-to-charge ratio and the specific aperture pattern. The spectrum can be computationally inferred from the spatial distribution at the detector plane. Resolution depends on the feature size in the coded aperture, whereas throughput depends on the number of features in the coded aperture.

A team from Duke University and RTI International used a 90-degree magnetic sector mass spectrometer to demonstrate a 10-fold increase in throughput with no loss in resolution by using a coded aperture miniature mass spectrometer (CAMMS; Fig. 4.5).[16] Subsequently, the team demonstrated the compatibility of spatial coding with other sector mass analyzer geometries.[17] Currently, the team is working on CAMMS-ES (coded aperture miniature mass spectrometer for environmental sensing) under funding from the ARPA-E MONITOR (US Advanced Research Projects Agency−Energy Methane Observation Networks with Innovative Technology to Obtain Reductions) program to combine the miniaturization-enabling

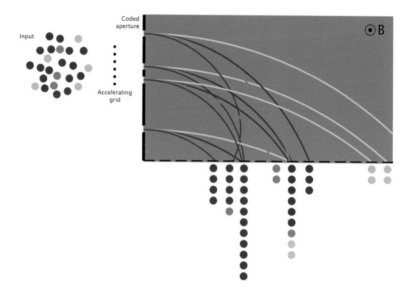

Figure 4.4 *Coded aperture mass spectrometer. A given location on the detector plate will receive a multiplicity of ions, and the spectrum is computationally inferred. The multiple apertures allow greater throughput.* Courtesy: Duke University.

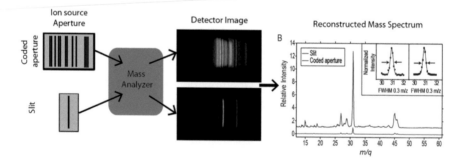

Figure 4.5 *Coded aperture improves resolution by an order of magnitude.* Courtesy: Duke University.

technology of aperture coding with several other miniaturization-enabling technologies, including microfabricated carbon nanotube field emission ion sources,[18] a cycloidal mass analyzer,[19] and a unique ion array detector developed by the University of Arizona.[20] The spectrum on the right in Fig. 4.5 demonstrates the efficacy of the coded aperture approach. Note the order of magnitude increase in intensity over the peak detected by a single slit. This is the most critical enabler of the shoebox-size mass spectrometer with resolution comparable to that of a laboratory based machine.

A mass spectrometer is the most versatile device available for deciphering the presence of potentially harmful airborne species. In the shale oil and gas industry, much of the reporting of health-related outcomes has been anecdotal, along with a few properly designed studies, but sound epidemiology will require accurate measurements to truly infer causality. For example, a study of 9384 mothers linked to 10,946 neonates who lived in a shale gas area found two preterm births but not low birth weight.[21] On the other side of the debate is a report that examined 15,451 births that occurred between 2007 and 2010, also in Pennsylvania, and that concluded a higher incidence of low-birth-weight infants but did not infer causality; the authors recommended more studies.[22] Not surprisingly, there was a rebuttal, and both sides are in the news item by Mills.[23]

Better regulations requiring the information on truck traffic associated with oil and gas operations and better sensing methods, such as those discussed in this chapter, could enable epidemiologic studies to illuminate the true hazards associated with shale oil and gas production. When the hazards are identified, amelioration is probably feasible. However, without quality studies, the industry will not be motivated to seek changes to practices.

The emphasis in this book is on fitness of purpose. However, scientific leaps in achieving such a prosaic goal are nevertheless cause for felicitation. They also often give the inventor a unique position in commercial exploitation. Advances in science are sprinkled through this book, interspersed with the more practical engineering to balance other features, especially cost.

One final point: Portable measurements of PM and VOCs could enable citizen science, which has been hampered, perceptually and really, by inadequate methodology. Such enablement would be greatly facilitated if the training requirement for instrument use were made minimal and if, importantly, the results were to be transmitted automatically to the cloud for retrieval by the state environmental authority.

REFERENCES

1. MacDonnell M. et al. *Mobile sensors and applications for air pollutants.* <https://cfpub.epa.gov/si/si_public_record_report.cfm?dirEntryId=273979&simpleSearch=1&searchAll=mobile+sensors+and+applications+for+air+pollutants>; 2013 [accessed 22.06.16].

2. US Office of the Director of National Intelligence. The MAEGLIN Program. <https://www.iarpa.gov/images/files/programs/maeglin/MAEGLIN_Proposers_Day.pdf>; 2016 [accessed 22.06.16].

3. Sarnat J, Wilson W, Strand M, Brook J, Wyzga R, Lumley T. Panel discussion review: session 1—exposure assessment and related errors in air pollution epidemiologic studies. *J Exposure Sci Environ Epidemiol* 2007;**17**(Suppl. 2):S75−82.

4. Rodes C, Lawless P, Thornburg J, Williams R, Croghan C. DEARS particulation relationships for personal, indoor, outdoor, and central site settings for a general population. *Atmos Environ* 2010;**44**:1386−99.

5. Weis B, Balshaw D, Barr J, Brown D, Ellisman M, Lioy P, et al. Personalized exposure assessment: promising approaches for human environmental health research. *Environ Health Perspect* 2005;**113**(7):840−8.

6. Schmidt C. Monitoring environmental exposures: now it's personal. *Environ Health Perspect* 2006;**114**(9):A528−34.

7. Amaral SS, de Carvalho Jr JA, Martins Costa MA, Pinheiro C. An overview of particulate matter measurement instruments. *Atmosphere* 2015;**6**:1327−45.

8. World Health Organization. *Burden of disease from ambient air pollution for 2012.* <http://www.who.int/phe/health_topics/outdoorair/databases/FINAL_HAP_AAP_BoD_24March2014.pdf>; 2014 [accessed 26.06.16].

9. Kim J.Y. Statement in interview. <https://search.yahoo.com/search;_ylc=X3oDMTFiN25l aTRvBF9TAzIwMjM1MzgwNzUEaXRjAzEEc2VjA3NyY2hfcWEc2xrA3NyY2h3ZWI-?p= world-bank-urges-better-cookstoves-developing-states-curb-&fr=yfp-t&fp=1&toggle=1&cop= mss&ei=UTF-8>; 2013 [accessed 24.06.16].

10. Loomis D, Grosse Y, Lauby-Secretan B, Ghissassi FE, Bouvard V, Benbrahim-Tallaa L, et al. The carcinogenicity of outdoor air pollution. *Lancet Oncol* 2013;**14**:1262−3.

11. Chartier R, Phillips M, Mosquin P, Elledge M, Bronstein K, Nandasena S, et al. A comparative study of human exposures to household air pollution from commonly used cook stoves in Sri Lanka. Indoor Air. 2016; January 21.

12. US Environmental Protection Agency Refinery Rules. <https://www.gpo.gov/fdsys/pkg/FR-2016-02-09/pdf/2016-02306.pdf>; 2016 [accessed 24.06.16].

13. Brunnee C. The ideal mass analyzer: fact or fiction? *Int J Mass Spectrom Ion Processes* 1987;**76**:125−237.

14. Ouyang Z, Cooks RG. Miniature mass spectrometers. *Ann Rev Anal Chem* 2009;**2**(1):187−214.

15. Muccio Z, Jackson GP. Isotope ratio mass spectrometry. *Analyst* 2009;**134**(2):213−22.

16. Chen EX, Russell ZE, Amsden JJ, Wolter SD, Danell RM, Parker CB, et al. Order of magnitude signal gain in magnetic sector mass spectrometry via aperture coding. *J Am Soc Mass Spectrom* 2015;**26**(9):1633−40.

17. Russell ZE, DiDona ST, Amsden JJ, Parker CB, Kibelka G, Gehm ME, et al. Compatibility of spatially coded apertures with a miniature Mattauch-Herzog mass spectrograph. *J Am Soc Mass Spectrom* 2016;**27**(4):578−84.

18. Natarajan S, Parker CB, Piascik JR, Gilchrist KH, Stoner BR, Glass JT. Analysis of 3-panel and 4-panel microscale ionization sources. *J Appl Phys* 2010;**107**(12):124508.

19. Bleakney W, Hipple Jr. JA. A new mass spectrometer with improved focusing properties. *Phys Rev* 1938;**53**(7):521−9.

20. Felton JA, Schilling GD, Ray SJ, Sperline RP, Denton MB, Barinaga CJ, et al. Evaluation of a fourth-generation focal plane camera for use in plasma-source mass spectrometry. *J Anal At Spectrom* 2011;**26**(2):300−4.

21. Casey JA, Savitz DA, Rasmussen SG, Ogburn EL, Pollak J, Mercer DG, et al. Unconventional natural gas development and birth outcomes in Pennsylvania, USA. *Epidemiology* 2016;**27**(2):163−72.

22. Stacy SL, Brink LL, Larkin JC, Sadovsky Y, Goldstein BD, Pitt BR, et al. Perinatal outcomes and unconventional natural gas operations in southwest Pennsylvania. *PLoS ONE* 2015;**10**(6):e0126425.

23. Mills D. Fracking may cause lower birth weights in babies, study says. *Healthline News.* <http://www.healthline.com/health-news/fracking-may-cause-lower-birth-weights-in-babies-060315#1>; 2015 [accessed 26.06.16].

PART II

The Potential for Contaminating Ground Water

Convictions are more dangerous enemies of truth than lies.
(Überzeugungen sind gefährlichere Feinde der Wahrheit, als Lügen.)
—Friedrich Nietzsche, Human, All Too Human

CHAPTER 5

Methane in Groundwater

Groundwater contamination occurs in two forms: (1) by hydrocarbons, either gaseous or liquid, and (2) by the direct spill of chemicals used in the drilling and completion process, especially fracturing fluids. This chapter discusses the gaseous element. It also discusses the operational basics, which apply to the liquid issue taken up in Chapter 6.

Fig. 5.1 is a schematic of a horizontal well. Although not to scale, the depths shown are fairly typical of shale plays. Freshwater aquifers rarely extend beyond 1000 feet and usually are shallower than 500 feet. Virtually all oil and gas wells have this profile. It is characterized by at least two sets of steel casing and associated cement between the fresh water and the inside bore of the well. The surface casing is required by state regulations to be a minimum depth below the lowest point of the aquifer. That distance varies between 50 and 100 feet. In better practices, it is as much as 300 feet. If aquifer contamination occurs from the bore hole, a flawed cement job is the likely culprit. If best practices are followed, this is extremely unlikely, although it is possible.

Fracturing is done in the horizontal section. In the schematic, one can observe that the closest approach of the aquifer is approximately 6000 feet. Microseismic-aided studies in the Barnett Shale have concluded that fractures rarely extend beyond 1500 feet from the horizontal borehole, as can be seen in Fig. 5.2. Each fracturing step is a small seismic event. Three-axis accelerometers detect the signals from the events, and successive measurements enable mapping of the fracture paths. The operational purpose of such measurements is to determine how much of the reservoir is being penetrated. The existence of natural fractures is a factor because the induced fractures interact with them. This is discussed in greater detail in Chapter 8.

Fig. 5.2 shows that individual fractures are variable in the vertical reach. This understanding is important for horizontal well placement

Sustainable Shale Oil and Gas. DOI: http://dx.doi.org/10.1016/B978-0-12-810389-0.00005-X

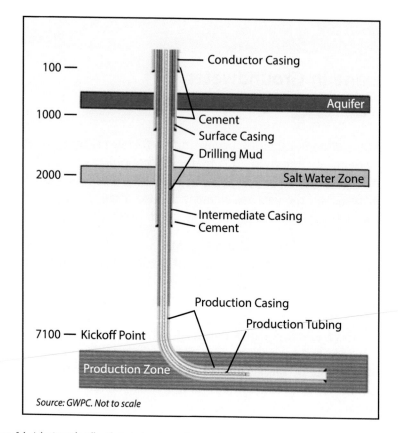

Figure 5.1 A horizontal well with typical casing and cementing practice. Not to scale. Courtesy: Groundwater Protection Council, further courtesy ALL Consulting.

because closely spaced laterals could well be in communication, which is not a desired result. For this chapter, the interest is in the possibility of fluids communicating with freshwater aquifers directly from the fracturing zones. At vertical distances of 2000 feet or more, this is very unlikely. Note, however, that if vertical natural fractures or faults exist, the microseismic measurement might not detect them because the sensors measure only the seismic event generated by energy transfer to the natural fractures from the induced fracturing process. If such transfer does not happen, those natural fractures will not be "seen" by the microseismic map—a factor in the possibility of induced earthquakes. Although not a subject for this book, the following box details the phenomenon and the analytical techniques that can help avoid it.

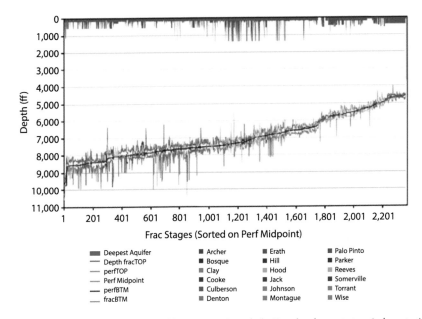

Figure 5.2 Fracture propagation as measured by microseismic methods. Note that the greatest vertical penetration toward freshwater aquifers does not exceed 1500 feet. Courtesy: American Oil and Gas Reporter.

Detecting and Avoiding Associated Earthquakes with Shale Oil and Gas Production

When sufficient energy is applied to an active fault, a seismic event takes place, which we know as an earthquake. A fault is a mismatch in adjacent rock, usually caused by an earlier geological event. An active fault is one in which slip occurs in the adjacent planes of rock in response to the input energy. If the slip is large enough, the result is an earthquake. According to the US Geological Survey, the amplitude of an earthquake is directly proportional to the length of the active fault involved.

These factors inform the measurements needed to assess the likelihood of an earthquake in response to a hydraulic fracturing operation. Active faults may be detected by three-dimensional seismic measurements. The technique launches a sound wave from the surface. The wave interacts with formations according to the equation

$$Z = V_\rho$$

where Z is the impedance of the rock layer being penetrated, ρ is the density of the rock, and V is the acoustic velocity. A portion of the wave is reflected and the rest is refracted through the rock. The relative portions are determined by the impedance. The refracted wave proceeds to the next rock layer, and once again a portion is reflected. All the reflected waves are detected on multiple sensors, which are usually on the surface. Analysis of the velocities yields a picture of the subsurface. This picture will show the faults, if any. All but the smallest faults will be detected. Because the magnitude of the potential seismic event depends on the length of the fault, small undetected faults are of no importance to the exercise.

In most instances of development of oil and gas fields, three-dimensional seismic measurements are done to determine the most productive locations in which to drill. Thus, fault detection is essentially automatic, and suspect locations are avoided. The potential for earthquakes from wells used for disposal of wastewater generated from fracturing operations is a different matter and is taken up in Chapter 6.

POTENTIAL FOR GAS MIGRATION

We have established that the likelihood of gas migration from the fracturing event is small. Measures are taken during well construction to ensure the integrity of the well, which is essentially a giant pressure vessel. The well casings shown in Fig. 5.1 are designed to withstand formation stresses, and breach is extremely unlikely. Improper cementing can be responsible for gas leakage. Cement formulations are designed for the environmental conditions and tested prior to injection into the borehole. The cement is typically injected down the steel pipe, emerges at the bottom, and rises up the annulus, bonding to both the casing and the formation as it rises.

A leak path for gas can be created if the cement does not bond adequately to the casing or the formation. In many instances, the leakage is described as channeling. The measurement to detect such a possibility is known as the cement bond log. Multiple acoustic transmitters and receivers are used in arrays. The waves penetrate the casing, cement, and formation rock, each of which has a different impedance. The absence of a bond between the cement and either the casing or the formation changes the impedance seen by the wave. Evaluation of the return waves provides the location and size of any gaps. The cement bond log was famously not run on the Deep Water Horizon well that blew out in the Gulf of Mexico in 2010. The reasons for waiving this routine operation are not clear.

The strength of the cement bond at the casing shoe is tested by a method known as the formation integrity test. The shoe is a special rounded profile at the very bottom of the casing. The test involves making the cased and cemented hole a pressure vessel by closing all release mechanisms such as the blowout preventers. Hydraulic pressure is applied until the open hole below the shoe fractures in a controlled manner. Analysis of the pressure transient yields information on the fracture gradient in the formation (the pressure at which fracture initiation will occur). The test also determines whether the cement around the casing shoe is competent.

Several types of formation naturally allow migration of methane to the surface through faults, fissures, and fractures. In the United States, this is most common in the East. The Eternal Flame Falls in the Shale Creek Preserve in New York State is a well-known location at which fugitive methane is kept lit. Aquifers with methane are common in Pennsylvania, which confounds the investigation of methane contamination attributed to gas wells.[1]

DISTINGUISHING BETWEEN SOURCES OF METHANE

Geoscientists have been dealing with this problem for several decades. There are two fundamental mechanisms of hydrocarbon production: thermogenic and biogenic. In the most simplistic sense, fossil fuel methane is thermogenic, and terrestrial methane, especially from organic waste and the rumens of animals, is biogenic. (There are minor exceptions.) The problem is addressed primarily by the four methods discussed next.

Carbon Isotope Monitoring

Carbonaceous material has a distribution of two isotopes of carbon, ^{12}C and ^{13}C. These are designated, respectively, light and heavy. ^{13}C has an extra neutron in its nucleus, hence the extra weight. Both are nonradioactive, with the property of radioactivity being reserved for ^{14}C, which is not discussed here. The lighter element has a greater vibrational frequency and consequently is more lightly bound. This makes it more reactive and gets it preferentially chosen for the reaction.[2] Incidentally, this applies also to the oxidation of methane by bacteria, which is discussed later. There is also a belief that migration of the fluids does not materially change the isotopic ratios of hydrocarbons.[3] This could be important in identifying the relative provenance of two possible reservoirs. Without getting ahead of ourselves, the dispute sometimes is whether the gas came from the reservoir that was fractured or from an intermediate horizon due to cementing inadequacies. Also, it is known that reservoirs are often fed from deeper sources of hydrocarbons (usually gas, not oil).

Carbon isotope ratios are expressed as $\delta^{13}C-CH_4$ in parts per thousand (denoted as ‰) deviations from a standard, which is a marine carbonate with a well-accepted ^{13}C:^{12}C ratio of 0.0112372. The comparison in all cases carries a negative sign because all known hydrocarbons have ratios lower than this standard.

The thermogenic mechanism, as the name implies, is from the action of heat (and pressure). As discussed in Chapter 2 methane is the most thermally mature state of the decomposition of kerogen. The isotopic ratio of the beginning material, in this case kerogen, determines the ratio in the final product. Consequently, each reservoir will be different, which can help to distinguish the provenance of the methane were some to migrate to a freshwater aquifer.[4]

Biogenic methane is sometimes termed bacteriogenic because it is formed as a result of metabolic processes in which CO_2, hydrogen, and acetate are converted to methane and water to produce energy. A well-known source is the rumens of animals, where the methane emerges as flatulence and burping. This represents an inefficient use of the cellulosic feed and in the United States comprises approximately one-third of all methane emissions. Another relevant source is bacterial action in freshwater aquifers. Bacterial formation of methane causes a greater fractionation than would a thermogenic mechanism, in part because bacteria strongly favor lighter isotopes. The other reason why biogenic gas has a lighter ratio is that the other reactants are themselves often products of bacterial action and so already have a built-in fractionated state.

The foregoing appears to provide a clear-cut means to differentiate origins. Biogenic methane is generally lighter, with $\delta^{13}C-CH_4$ approximately −90‰ to −55‰. Thermogenic methane is heavier, with $\delta^{13}C-CH_4$ approximately −55‰ to −35‰. Now the confounding factors: mixing and oxidation. Mixing refers to thermogenic gas getting diluted by gas from another source with different isotopic character.

Oxidation refers to the action of bacteria to oxidize a portion of the methane. The implications are discussed next.

Attributing stray gas to a source formation requires consideration of both the source signature, representing a predominantly thermogenic gas in the deep formation, and mixing with shallow biogenic methane that forms in aquifers in which drinking water wells are drilled. The carbon isotopic signature of methane, $\delta^{13}C-CH_4$, varies between Appalachian Basin gases, depending on the age and thermal maturity of the formation, with the more positive values (approximately $-30\permil$) seen in Marcellus shale and more negative values seen in other formations. However, the more negative values can be observed when deep, thermogenic gas mixes with shallow biogenic gas in aquifers, as can occur in a stray gas scenario. Indeed, stray gas incidents have been shown to exhibit an increase in the thermogenic portion of methane.[5] As noted previously, because biogenic gas has a more negative $\delta^{13}C-CH_4$ than thermogenic gas, $\delta^{13}C-CH_4$ in the mixing trend between Marcellus shale gas and shallow biogenic gas overlaps the $\delta^{13}C-CH_4$ range of thermogenic gases from rock of other ages in the basin. Therefore, $\delta^{13}C-CH_4$ used alone may not clearly delineate the source formation. One would not be in a position to attribute the gas to the zone being fractured, and in fact the mid-range ratio may erroneously point to a completely different formation. As a practical matter, the remedy may not be much different. No matter the provenance of the gas, the mechanism for migration to the aquifer is most likely improper well sealing in the vertical portion of the well.[5–7] The fix would be the proper sealing of that portion of the well.

Another confounding factor is bacterial oxidation of methane. Just as in methane production, the bacterial oxidation process prefers the lighter isotopes, leaving behind a residual methane that is now heavier. The presence of such bacteria can complicate the inferences from a simple $\delta^{13}C-CH_4$-based interpretation.

A more specialized technique of fingerprinting gas sources uses $\delta^{13}C-CH_4$ in conjunction with $\delta^{13}C$ of ethane ($\delta^{13}C-C_2H_6$). Marcellus shale gases show a distinctive pattern in which $\delta^{13}C-CH_4$ is more positive than $\delta^{13}C-C_2H_6$, unlike the thermogenic gases in overlying formations, in which $\delta^{13}C-CH_4$ is more negative than $\delta^{13}C-C_2H_6$.[8,9] Studies suggest that certain noble gas isotopes (e.g., ^4He) identify specific thermogenic gas sources.[5,10] They benefit from the fact that they are unaffected by microbial action such as the oxidation of methane, which confounds the other isotopic measurement technique.

Hydrogen Isotope Monitoring

Although the carbon isotope appears to be the most useful differentiator, hydrogen can be used as well. Hydrogen with two neutrons, deuterium, is the heavy isotope. The δ^2H of thermogenic methane depends on the value in the original source material. This can vary significantly even within a shale basin. The δ^2H of biogenic methane is similarly related to the source materials, here H_2O, H_2, and acetate. The water is often seen as a differentiator because fresh water, the source for much biogenic gas, results from precipitation, during which process strong fractionation of hydrogen occurs. Later in the chapter, we illustrate the use of a cross-plot of hydrogen fractionation with that of methane to identify the likely rock layer that produced the gas in question.

Monitoring C_2+ Species

Because of some limitations of the hydrocarbon isotopic method, investigators have identified another independent variable. They have shown that the nonmethane gases in a sample can effectively fingerprint the shale source. For example, higher chain hydrocarbons such as ethane (C_2H_6) and propane (C_3H_8) are present in thermogenic gas but virtually absent in biogenic gas. Chapter 2 explained why this finding should apply to most shale gas. The ratio of methane to higher chain hydrocarbon abundance ranges from less than 10 to greater than 10,000, providing a second axis that has been used in combination with $\delta^{13}C-CH_4$ to identify various thermogenic and biogenic gas sources and mixing trends among these sources. The plot of the $\delta^{13}C-CH_4$ value versus the ratio of methane to higher chain hydrocarbon has been widely used in studies of natural gas[11] and in the Appalachian Basin (e.g., Osborn et al.[1]). Similarly, $\delta^{13}C-C_2H_6$ can be plotted against the ratio of methane to larger hydrocarbons, most commonly methane/ethane.[12] Fig. 5.3 shows one such cross-plot to illustrate the technique.

Fig. 5.3 is an example of a clean separation of biogenic and thermogenic gas origins. The thermogenic quadrant on the lower right is characterized by a relatively high proportion of C_2+ (ethane or larger) molecules compared to the methane, yielding a low ratio of $C_1:C_2+$.

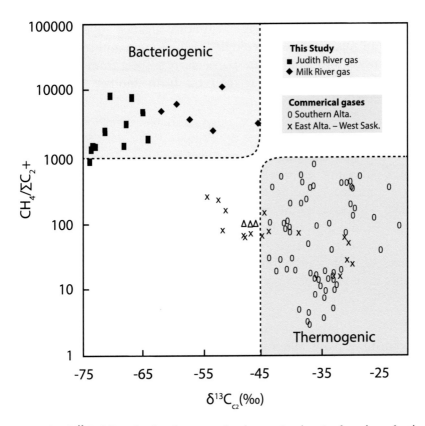

Figure 5.3 Plot of $\delta^{13}C-C_2H_6$, rather than the more usual methane, against the ratio of prevalence of methane and larger molecules. Taylor SW, Sherwood Lollar B, Wassenaar LI. Bacteriogenic ethane in near-surface aquifers: implications for leaking hydrocarbon well bores. Environ Sci Technol 2000; **34**: 4727−32.

At the same time, the $\delta^{13}C-C_2H_6$ value is less negative (heavier). This is a standard description of thermogenic gas without confusing factors such as mixing. In contrast, the biogenic gas in the upper left quadrant is characterized by a high $C_1:C_2+$ ratio combined with a more negative $\delta^{13}C-C_2H_6$ (lighter) value. This is standard biogenic gas.

Cross-plots of $C_1:C_2+$, $\delta^{13}C-CH_4$, and δ^2H-CH_4 were used in all the studies. In some of the references, the $C_1:C_2+$ is referred to as "wetness" because the larger molecules are mostly liquid at normal temperatures. In industrial parlance, this compendium is called natural gas liquids. $\delta^{13}C-CH_4$ and δ^2H-CH_4 cross-plots have been used to identify the age of the rock associated with the methane, and to some extent there is an orderly progression along the age of the rock,[13] which some have termed a thermal maturation pathway.[14] Sprinkled through most references are cautions against overinterpretation because of mixing issues. Geologies prone to non-anthropogenic methane intrusion into aquifers are particularly suspect from the standpoint of mixing. "Charging" of a reservoir from another stratum is not uncommon.

Coal Seam Gas

Coal seam gas, also known as coal bed methane, deserves a discussion here. Coal seams are often encountered while drilling for oil and gas. Good practice requires that these zones be properly sealed with cement. However, to the extent that they can be implicated in methane intrusions into aquifers, an understanding of the geochemistry is necessary. Furthermore, coal bed methane is an important hydrocarbon source. Accidents at coal mines in proximity to natural gas fields may require an understanding of whether methane infiltrated the coal mines from other sources.

Coal was formed through geochemical and biochemical reactions transforming plant material into carbon-rich solids. Further thermal- and pressure-induced maturation converted some of it into methane. There can be low single-digit percentages of ethane but virtually no larger hydrocarbons such as propane or butane.[15] In this aspect, coal seam gas is more similar to biogenic than thermogenic gas. However, the carbon isotope story is quite different. Central Appalachian Basin sourced coal bed methane indicates $\delta^{13}C$ values for methane of $-39.9‰$ to $-55.1‰$[16] which is in the same range as petroleum-based methane, not entirely surprising considering the similarities in the creation mechanisms. A suspected wellbore leak would need to rely primarily on the $C_1:C_2+$ ratio to eliminate a coal layer as the source. That would also be the yardstick for determining the infiltration of petroleum sourced methane into coal mines.

INVESTIGATIONS OF METHANE MIGRATION TO AQUIFERS

Gas migration has been extensively investigated. The most prominent investigation targeting shale oil and gas operations is that of Osborn et al.[1] They studied several water wells in proximity to active and inactive shale gas production operations. All of the active gas wells were in Susquehanna County, Pennsylvania, and were producing from the Marcellus formation. The methane in water wells was strikingly inversely correlated with distance from active wells. Isotopic studies of the sort described previously showed that the methane was of thermogenic origin. They further concluded that the methane was from rock of an age consistent with the Marcellus formation. Importantly, they found no liquid contaminants from fracturing fluid.

Osborn et al.'s study predictably created an international stir because it was the first to report evidence of water contamination from shale gas wells. Although causality with hydraulic fracturing was inferred, especially in the title of the paper, "Methane Contamination of Drinking Water Accompanying Gas-Well Drilling *and Hydraulic Fracturing*" (emphasis added), none was established. In part, that was because the test area was known to have non-anthropogenic gas migration to water wells, and no baseline measurements were taken prior to the drilling operations. However, the isotopic studies demonstrating the inverse correlation of the contamination with distance from the wells were compelling. A possible explanation remained that improper well construction in the vertical sections, and not the fracturing zones, caused the leaks. This is generally acknowledged as a viable risk[5,17] in any gas production, with or without hydraulic fracturing. The Deep Water Horizon blowout is believed to have originated from such a cause. Most states now have regulations to avoid breaches of the wellbore.[18]

Predictably, the Osborn et al. paper was followed in the same year by Molofsky et al. rebutting the findings.[14] They reexamined the Osborn data in addition to providing independent observations. One of their key points is that natural upward migration of methane is common in the area. They demonstrate a strong correlation of water well contamination with topography (Fig. 5.4). In particular, the incidences are greater in valleys. They contend that studies of contamination proximal to drill sites should normalize for this parameter. The most technically interesting aspect of the Molofsky et al. paper is the attribution of the source of methane to rock of a certain age.

Fig. 5.4 illustrates the use of the $\delta^{13}C-CH_4$ and δ^2H-CH_4 cross-plot. The maturation pathway appears fairly well behaved in following a near-linear path from younger to older rock. The Marcellus is shown in the upper right as distinct from the Middle Devonian, even though in precise terms it is part of that epoch. The Upper Devonian is in the range 359−385 million years, whereas the Middle Devonian is

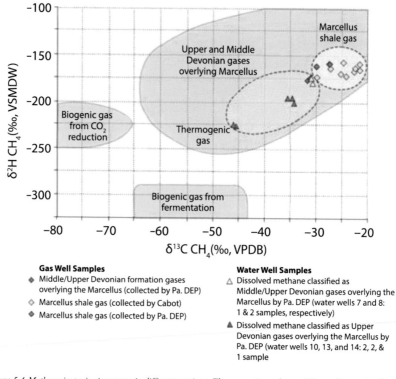

Figure 5.4 Methane isotopic signatures in different settings. Thermogenic gas has a different fingerprint depending on the age of the rock in which it resides. Courtesy: Molofsky L, Connor J, Farhat S, Wylie A Jr, Wagner T. Methane in Pennsylvania water wells unrelated to Marcellus shale fracturing. Oil Gas J 2011;109(49):54–67.

385–398 million years. Several water samples with gas in them were collected, as were formation gas samples. These samples were collected by the Pennsylvania Department of Environmental Protection, Cabot Oil, and some of the authors.

Molofsky et al.[14] claim to have taken the data from the study by Osborn et al.[1] and show them to fall squarely in the Upper and Middle Devonian zone, as opposed to the Marcellus. They conclude that the Osborn gas should be attributed to the younger epoch, not the Marcellus. Consider, however, that the separation here is by just a few tens of millions of years, which in geologic terms is short. The larger question is whether the gas came from the wellbore or occurred naturally. It remains unanswered, and as previously noted, it may not be highly material because remedies are available. The more dangerous actor is liquid contamination, which is the subject of Chapter 6.

Other studies have been conducted since that of Molofsky et al. some with baseline measurements. The consensus is that if gas migration happens, the cause is faulty wellbore construction and not the fracturing process. Any oil and gas operation could be subject to this flaw. The remedies include better measurements of fugitive emissions and stricter regulations governing well construction.

Methane per se in drinking water is not toxic at the levels observed. However, many view methane contamination as something of a canary in a coal mine: If it leaks from a well, surely bad actor chemicals from fracturing fluid could not be far behind. Because gaseous fluid can traverse strata more readily than liquids, methane as a harbinger of worse chemicals is not necessarily a sound argument.

Existing analytical tools adequately distinguish between biogenic and thermogenic sources. The precise geologic origin of the thermogenic source is still somewhat speculative. In the view of this author (VR), the trends are not well behaved enough to be unequivocal, especially when the competing strata overlay each other in geological time (see Fig. 5.4 and the associated discussion). If baseline testing is done to determine the existence of methane in water wells prior to drilling petroleum wells, much of the uncertainty goes away. Then, if at some stage methane is detected, one would be well positioned with current analytical techniques to ascribe provenance.

REFERENCES

1. Osborn SG, Vengosh A, Warner NR, Jackson RB. Methane contamination of drinking water accompanying gas-well drilling and hydraulic fracturing. *Proc Natl Acad Sci* 2011;**108**:8172−6.

2. Krauskopf KB, Bird DK. *Introduction to Geochemistry*. 3rd ed. Boston: McGraw Hill; 1995 [647 p].

3. Schoell M. Genetic characterization of natural gases. *Am Assoc Pet Geol Bull* 1983;**67**:2225−38.

4. Breen KJ, Revesz K, Baldassare FJ, McAuley SD. Natural gases in ground water near Tioga Junction, Tioga County, North-Central Pennsylvania—Occurrence and Use of Isotopes to Determine Origins, 2005. USGS Scientific Investigations Report 2207-5085; 2007.

5. Darrah TH, Vengosh A, Jackson RB, Warner NR, Poreda RJ. Noble gases identify the mechanisms of fugitive gas contamination in drinking-water wells overlying the Marcellus and Barnett shales. *Proc Natl Acad Sci* 2014;**111**:14076−81.

6. Kresse TM, Warner NR, Hays PD, Down A, Vengosh A, Jackson RB. Shallow groundwater quality and geochemistry in the fayetteville shale gas-production area, North-Central Arkansas, 2011. US Geological Survey Scientific Investigations Report 2012-5273. <http://pubs.usgs.gov/sir/2012/5273/>; 2012 [31 p] [accessed 14.07.16].

7. King GE. *Hydraulic Fracturing 101: What Every Representative, Environmentalist, Regulator, Reporter, Investor, University Researcher, Neighbor and Engineer Should Know About Estimating Frac Risk and Improving Frac Performance in Unconventional Gas and Oil Wells. SPE 152596-MS*. The Woodlands, TX: Society of Petroleum Engineers; <http://www.onepetro.org>; 2012 [accessed 19.07.16].

8. Osborn SG, McIntosh JC. Chemical and isotopic tracers of the contribution of microbial gas in Devonian organic-rich shales and reservoir sandstones, northern Appalachian Basin. *Appl Geochem* 2010;**25**:456–71.

9. Jackson RB, Vengosh A, Darrah TH, Warner NR, Down A, Poreda RJ, et al. Increased stray gas abundance in a subset of drinking water wells near Marcellus shale gas extraction. *Proc Natl Acad Sci* 2013;**110**:11250–5.

10. Hunt AG, Darrah TH, Poreda RJ. Determining the source and genetic fingerprint of natural gases using noble gas geochemistry: A Northern Appalachian Basin case study. *Am Assoc Pet Geol Bull* 2012;**96**:1785–811.

11. Bernard BB, Brooks JM, Sackett WM. Natural gas seepage in the Gulf of Mexico. *Earth Planet Sci Lett* 1976;**31**:48–54.

12. Taylor SW, Sherwood Lollar B, Wassenaar LI. Bacteriogenic ethane in near-surface aquifers: implications for leaking hydrocarbon well bores. *Environ Sci Technol* 2000;**34**:4727–32.

13. Jenden PD, Drazan DJ, Kaplan IR. Mixing of thermogenic natural gases in northern Appalachian Basin. *Am Assoc Pet Geol Bull* 1993;**77**:980–98.

14. Molofsky L, Connor J, Farhat S, Wylie Jr A, Wagner T. Methane in Pennsylvania water wells unrelated to Marcellus shale fracturing. *Oil Gas J* 2011;**109**(49):54–67.

15. Kim AG. *The composition of coalbed gas. Report of investigations 7762*. Washington, DC: US Bureau of Mines; 1973 [9 pages].

16. Laughrey CD, Baldassare FJ. Geochemistry and origin of some natural gases in the plateau province, central Appalachian basin, Pennsylvania and Ohio. *Am Assoc Pet Geol Bull* 1998;**82**:317–35.

17. King GE. Thirty years of shale gas fracturing: what have we learned? *SPE 133456-MS*. The Woodlands, TX: Society of Petroleum Engineers; <http://www.onepetro.org>; 2010 [accessed 14.03.12].

18. Rao V. *Shale oil and gas: the promise and the peril*. 2nd ed Research Triangle Park, NC: RTI Press; 2015.

CHAPTER 6

Potential for Liquid Contamination of Groundwater

Liquid contamination of groundwater is possibly the public's largest concern with shale oil and gas production. The concern centers primarily on the constituents of fracturing fluid, partly premised on the belief that the fluid contains carcinogens. This chapter describes the constituents of fracturing fluids and the progress that has been made, largely through regulation, to limit harmful chemicals in their composition. It also discusses the studies to date on assessing the risk.

When fracturing fluid is injected into the rock, approximately 16 to 35% returns in what is known as flowback water. There are instances of higher and lower percentages, but this is a common range.[1] Most of the injected water remains in the formation. Concern has been raised about the fate of this water. In particular, theories are advanced to suggest that it could migrate up to freshwater aquifers over time through high-permeability vertical fractures.[2] A recent study examined the possible mechanisms, conducted discrete experiments, and concluded that such migration is very unlikely.[3] In our opinion, water residues left behind in the formation should be considered sequestered.

FLOWBACK WATER

Because the water in petroleum reservoirs is very saline, the flowback water is always more salty than the water that went in, if fresh water is used at the outset.[4] The salinity of formation water varies, but in general it increases with depth, content of paleo-seawater in the formation, and age of the rock. Consequently, flowback water will have salinity varying from 16,000 to 300,000 ppm (1.6 to 30%). In addition, it will have some fraction of the injected chemicals, usually less than 10% of the injected volume.[5] By and large, these chemicals are consumed (e.g., biocides and polymers) or simply left behind in the formation.

Sustainable Shale Oil and Gas. DOI: http://dx.doi.org/10.1016/B978-0-12-810389-0.00006-1

The following are the principal chemicals in injected fracturing fluid:

- Gelling agent: The agent is guar or xanthan gum or hydroxyethyl cellulose. Guar, which is by far the most used of these agents, is a polymerized disaccharide of mannose and galactose. This is rarely used for shale gas but routinely used for shale oil or when there is high condensate cut in the shale gas. It increases viscosity to allow suspension of the proppant, which is often sand, with specific gravity of approximately 2.65. Lighter synthetic proppant materials that can be carried in water without a gelling agent are under development. Guar in much more purified form is used to thicken ice cream.
- Cross-linker: This is employed to cross-link the polymer, which is usually a derivative of guar such as carboxymethyl hydroxypropyl guar. Cross-linking increases the viscosity over the simple polymer. Borates are most common because the borate-linked gel is also shear thinning (lower viscosity at higher shear). However, a zirconate cross-linker has been found to be effective up to salinities of 285,000 ppm[6] and likely beyond. This discovery was a key aid to the important practice of reuse of flowback water, which is discussed later.
- Breaker: This breaks down the cross-links to enable the carrier fluid to be returned to the surface after the proppant has been deposited. Commonly used breakers are ammonium persulfate and magnesium peroxide. With the zirconium cross-linker, the breaker is sodium chlorite.
- Proppant: This is hard, inorganic material that props open the fractures to prevent closure by earth stresses. The most common proppant is quartz sand. For higher pressure zones, bauxite (oxide of aluminum) is used, as are synthetic ceramics.
- Friction reducer: Some agent is always used, usually a polyacrylamide but sometimes a petroleum distillate, which is increasingly forbidden by regulation. A reducer is particularly required in "slick water" fracturing formulations containing no gelling agent. Water has a high coefficient of friction, and a reducer is needed to minimize the energy for pumping. Polyacrylamide, which is highly water absorbent, is the most common reducer. It is also used in baby diapers and as a flocculent to remove fine particles in drinking water.

- Scale inhibitor: This is used in approximately 25% of cases, depending on the solutes present. It is more important if radioactive species are expected because they tend to concentrate in scale. Also, when flowback water is reused without removal of divalent ions, the likelihood of scaling increases. Phosphonate and ethylene glycol are most common. They are also constituents of household detergents and antifreeze, respectively.
- Biocide: This is used in almost every instance to control bacteria present in the reservoir, which potentially harm reservoir performance. The agent is glutaraldehyde or, less commonly, chlorine dioxide. Glutaraldehyde is a medical disinfectant, and chlorine is used in municipal water supplies. Increasingly, ultraviolet radiation at the surface is replacing biocide in the injected fluid.
- Surfactant: This is sometimes used to decrease the surface tension to allow water recovery. Many formulations use ethanol, isopropyl alcohol, or 2-butoxyethanol, which are also household products.
- Acid: Hydrochloric acid is used to dissolve mineral deposits and sometimes to initiate fractures in the rock.

The total concentration of these chemicals, not counting the inert ceramic proppants, rarely exceeds 0.5%. Most service companies now disclose the chemicals on their public websites.[7] A web link to Halliburton's fluids disclosure is an example,[8] which shows a map of the world with a searchable feature for the constituents of fracturing fluid at each location. Importantly, disclosure includes the Chemical Abstracts Service (CAS) Registry Number, from which all properties of interest, including toxicity, can be determined. Consequently, any controversy usually surrounds the constituents not disclosed because the user seeks intellectual property protection under the rubric of trade secrets. This is discussed in Chapter 10, which covers regulatory issues. In general, if one does not know what is in the fluid, one is not in a position to detect it in aquifers or elsewhere. The exception is the use of sentinel compounds, which are species that indicate flowback or produced water contamination. These can act as proxies for contamination without the need of testing for every species. According to one investigator, the most reliable sentinel compounds are total dissolved solids (TDS), chloride, and divalent cations.[9] In sampling 44 wells from four different plays, she found that barium and strontium correlated with TDS. TDS and chloride are logical sentinels because, as noted previously, flowback water invariably has high concentrations of them, certainly compared to fresh water (defined as $<500\,\mathrm{ppm}$ chloride) in aquifers. However, TDS and chloride are not specific to hydraulic fracturing fluid but, rather, are derived from the natural formation water. These salts could result from the flowback process or from natural migration of saline formation water.

The most recent regulations passed by states in the US ban the use of diesel and other aromatic-containing compounds in the fracturing water. They were used for friction reduction; polyacrylamide performs better. Accordingly, if BTEX compounds (benzene, toluene, ethylbenzene, and xylenes) are found in the flowback, they are probably naturally occurring. This is also the case for arsenic and radioactive minerals, which are typically naturally occurring where they occur in flowback waters, rather than being part of the chemicals used in hydraulic fracturing. In part due to this logic and in part due to the low concentrations generally found, volatile organic compounds (VOCs) are not good sentinels.[9] In no measure does the provenance mitigate the associated hazard.

DISPOSITION OF FLOWBACK AND PRODUCED WATER

Flowback water is defined as the water-based fluid returning to the surface immediately after the fracturing operation. Many days later, formation water will continue to add to the return water. At some stage, the fluid is predominantly formation water, and at that stage it is known as produced water. In operations in which no fracturing fluid is involved, all the water returning is produced water. This mixing relationship is responsible for the observation that flowback water becomes more salty with the passage of time.[1]

Fig. 6.1 is a schematic of the fate of flowback water. The source water is usually surface water or groundwater that is treated, if necessary. Fracturing fluid is formulated using this water as a base. The term slick water is used for formulations without any gel. This is commonly done for gas wells, following the observation that gel use was correlated with lower recovery, likely because of permeable pathways being clogged with residue. However, the treatment volume is greater in slick water operations. High flow rate is used to overcome the shortcoming of lower viscosity in the enabling of sand transport into the fractures.

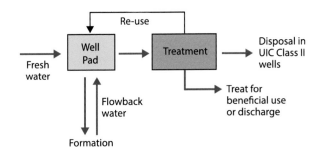

Figure 6.1 Disposition of flowback water.

The flowback water is sent for treatment. When it must be immediately reused on the next well, the treatment may be minor. Typically, solids are first removed by settling, which may be sufficient for slick water operations. If gel is to be used, electrocoagulation and filtering may be performed. Electrocoagulation in its simplest form is the application of radiofrequency energy to water, effecting a change in the surface charge state of the solids. In the modified form, the solids coagulate and the aggregates are large enough to be removed with mechanical filters.[10] Variants of the technique have been successful in removing heavy metals, the metalloid arsenic, and colloids.

In rare instances, divalent ions may be removed. The purpose is usually to minimize scale formation by carbonates of primarily calcium and magnesium. Also, if radioactive elements are present, they tend to segregate to the scale. Even if the concentrations in the water are so low as to not be a concern, the aggregation in scale can represent a hazard in scale removal, especially in surface tanks. Divalent ions may be removed through ion exchange mechanisms, such as in water softening. Electrocoagulation with membrane separation is also effective.

Some operators remove bacteria prior to reuse. Considering that the bacteria in question originated in the subsurface, returning them should be safe. However, there is recent evidence to support the notion that ingredients in fracturing fluid stimulate the growth of minority microbe populations resident in the reservoir (L. Ursell, personal communication of unpublished data, 2016). If this is confirmed, then perhaps removal is wise, if the newly energized species are deleterious. The generic objective of adding biocides to any fracturing fluid, whether fresh or reused, is to control the activity of sulfate-reducing species (heterotrophic bacteria in a substantially anaerobic environment) and the resultant production of hydrogen sulfide. Two methods are employed: chemical addition and ultraviolet radiation. The radiation is absorbed by the bacteria, and it damages the DNA to the point where protein synthesis and, more important, replication are interrupted. This process is ineffective in turbid water because of poor penetration of light. Turbidity can be removed beforehand by electro-coagulation and filtering.

Desalination is generally unnecessary. Even gel formulations are now able to tolerate salinities of up to 280,000 ppm.[6] This is economically and environmentally important because the need to desalinate down to fresh levels would discourage reuse.

The common alternative to reuse is disposal in injection wells. These carry a US Environmental Protection Agency (EPA) classification of Underground Injection Control (UIC) Class II and comprise injection into deep geologies, usually saline aquifers. Suspended solids are almost always removed. Injection is a very low-cost option, as little as US$ 0.25 per barrel. Its popularity has resulted in more than 150,000 such wells being drilled. Some of these are now known to have caused seismic activity.

Seismic activity created by fracturing has been a subject of debate. Certainly, each fracturing event generates seismic energy. That it does so is the basis for the investigative tool of microseismic monitoring described in Chapter 8. This passive listening method detects the acoustic signals and enables mapping of the location and progress of fractures. The fact that microseismic monitoring was common early on allowed investigators to quickly conclude that earthquakes directly related to hydraulic fracturing were extremely unlikely. The largest estimates of magnitude from such a source are in the vicinity of 2.5 to 3.0 on the Richter scale, which are well below the level of concern.

Wastewater injection into UIC Class II wells, on the other hand, is now definitely believed to create earthquakes, as reviewed in a US Geological Survey study.[11] Earthquakes attributed to wastewater disposal are believed to have generated magnitudes up to 5.1. This is a potentially damaging level and requires attention. The remedy would be to simply locate the disposal wells distant from such faults.[12] Also, each disposal event could be monitored with microseismic methods, and parameters could be developed for safe injection volume and pressure for each well. Regulations in this area are well overdue, as discussed in Chapter 10.

Many of the eastern and central portions of the Marcellus and Utica are geologically incompatible with deep disposal. Because reuse is the most inexpensive alternative, more than 90% of wells in Pennsylvania followed this practice in 2012,[13] and a higher percentage are likely doing so currently.

Flowback water not reused or pumped into injection wells can be treated for discharge or some other use. This is the most expensive option. The constituents to be removed are TDS up to 300,000 ppm, suspended solids, dispersed oil, VOCs, heavy metals, radionuclides, dissolved gases, bacteria, and fracturing fluid additives such as biocides, scale and corrosion inhibitors, gelling agents, and cross-link breakers.

Desalination is the most capital-intensive operation in treating flowback or produced water. The workhorse method is reverse osmosis, the simple principle of which is to cause water molecules to traverse a semipermeable membrane against the chemical potential gradient of water. This is achieved by applying pressure to the salty side to the point at which the chemical potential gradient is reversed and water molecules transport across the membrane to the freshwater side. Ordinarily, fresh water is at a higher level of chemical activity and will flow from the fresh to the salty side. However, as a practical matter, the energy consumed in pressurizing the salty side is too great to be economically feasible at a salinity level greater than approximately 70,000 ppm chloride. Consequently, this technique is used only for water up to approximately 40,000 ppm. The residual brine has to be disposed of, often into the ocean. Reverse osmosis has important applications, such as producing potable water from seawater ($\sim$ 35,000 ppm) and brackish groundwater (less than $\sim$ 10,000 ppm) in areas without much fresh water, such as the Middle East. However, for the highly saline waste of areas such as the Marcellus, it is not feasible.

Forward osmosis is an interesting technique and has greater potential because it uses less energy. A "draw" solution contains a chemical that produces a lower potential relative to the wastewater on the other side of the semipermeable membrane, causing water molecules to flow down the gradient to the solution side. The chemical is then removed, sometimes by volatilizing it for reuse, leaving behind clean water.[14] The heavy brine left on the wastewater side must be disposed of.

Membrane distillation is relatively new.[15] It uses a membrane that transmits only vapor by having a hydrophobic barrier on the wastewater side. Water vapor moves across, leaving behind the salts. VOCs, if any, must be removed first or they will move across also. The allure of this technique is that low-grade heat can be used to produce vapor. One innovative method combines it with forward osmosis.

A popular method is to simply evaporate the water from the waste side and condense it for use. It is energy intensive because the latent heat of evaporation must be provided. The resultant product is very pure. Some outfits claim to have developed techniques that minimize energy use. This type of technique may be needed for the very heavy brines that are too salty for direct reuse or reverse osmosis. However, evaporation does become more energy intensive as the salt content rises.

INVESTIGATIONS OF GROUNDWATER CONTAMINATION

Three potential sources for groundwater contamination have been investigated. The first is the direct communication from the fracturing event in the reservoir, through the now-enhanced fracture network and into the shallow subsurface. The second is leakage from the vertical portion of the well. The third is surface spill of the neat chemicals used in well operations. The analytical methods are similar, especially for the first two.

The most notable and documented case is that of Pavillion, Wyoming. After anecdotal observations of contamination of freshwater bodies, the EPA installed two monitoring wells in close proximity to suspected oil and gas sources. Since the initial EPA draft report of findings,[16] many investigators have examined the data.[17,18] The most recent analysis is by DiGiulio and Jackson.[19] Complicating the situation is the fact that coal bed methane production was also practiced in the area. The significance is that coal bed methane deposits tend to be very shallow and are often hydraulically fractured. Many of the fractured wells were as shallow as 230 meters, in some cases shallower than the deepest freshwater source (DiGiulio and Jackson, Figure 2a).[19] The surface casing in this area is reported to be in the range of 100 to 706 meters, with a median of 185 meters. This last figure is shallower than the deepest freshwater sources. To further compound the problem, no cementing was required below surface casing (and still is not) in that area.[18] Many instances of contaminants that could be attributed to the hydraulic fracturing or acidizing operations were found in well water.

All of the foregoing highlight the fact that cemented liners, the best line of defense against leakage, are critical. All recent state regulations in the United States require surface casing to extend more than 30 meters below the deepest freshwater layer. The liner below the surface casing is also required to be cemented.[20] Some of the studies discussed later were performed on wells such as these. Yet another complication in the Pavillion case was the earlier practice of allowing wastewater from fracturing to be placed in unlined pits. This too is not permitted under recent regulations. The North Carolina regulations,[20] for example, require double-lined pits with sensors between the liners to detect breach of the first liner. In summation, the Pavillion situation is not a proxy for what one should expect in more recent operations elsewhere with better regulations and associated practices.

In possibly the earliest detailed evaluation of fluid contamination of aquifers, methane was found in water wells proximal to shale gas wells. This finding was discussed in Chapter 5.[21] To estimate the potential for liquid contaminants, Osborne et al. examined expected constituents of flowback water. The Na^+, Ca^{2+}, and Cl^- ion concentrations were not significantly different from those of previous studies in water from the same formations. Because high salinity of flowback water is an especially distinguishing feature of these Marcellus plays, the ion concentrations are important sentinels.[9] The stable isotope signatures of $\delta^{18}O$ and δ^2H were also not significantly different. The isotope ratios for dissolved organic compounds, $\delta^{13}C-DIC$, were similarly negative. In short, every geochemical and isotopic measure of liquid contamination from wellbore fluids was negative.

A later study built on the one just discussed, with some of the same authors.[22] This added 109 newly collected shallow groundwater samples that, combined with the previous set, brought the total to 426. The analytical methods were similar, with high reliance on isotopic signatures in addition to solute species. Fig. 6.2 shows the results for δ^2H versus $\delta^{18}O$. As noted in the figure legend, the isotopic signatures of water in the aquifers follow an expected trend for freshwater bodies and do not approach the area reflective of the Marcellus shale. The authors also studied isotopes of strontium. They concluded that liquid contamination from oil and gas wells is unlikely. However, they noted evidence of natural pathways for upward communication, especially for gas. This is consistent with other observations of methane in

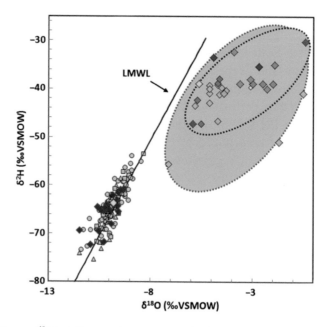

Figure 6.2 $\delta^2 H$ versus $\delta^{18}O$ in shallow groundwater near Marcellus shale gas wells and in Appalachian brines. The water isotope compositions of the shallow groundwater samples appear indistinguishable from each other and the local meteoric water line (LMWL)[23] and do not show any apparent trends toward the stable isotope ratios of the Appalachian brines (Devonian, Ordovician, and Silurian epochs) in the larger oval.[24] The smaller oval represents the Middle Devonian and older, reflective of the Marcellus shale. Courtesy: National Academy of Sciences.

aquifers unrelated to oil and gas operations. Because 3 D seismic investigation of the subsurface is most often done prior to exploitation, these conductive pathways should be detectable.

One recent study drew inferences to direct liquid contamination of freshwater wells from the fracturing activity in petroleum reservoirs.[25] This is contrary to a growing consensus that the connection is unlikely. Consequently, the paper drew much commentary. The authors followed up on incidences of foaming in certain water wells. They used a technique wherein water samples were analyzed by gas chromatographic separation using "GCxGC-TOFMS, isotope ratio mass spectrometry, and inductively coupled plasma atomic emission spectrometry (ICP-AES)." They identified a specific compound used as a surfactant in oil wells: 2-*n*-butoxyethanol. It was found in one of the foaming domestic water wells at nanogram-per-liter concentrations. Of course, this contamination could have originated from a poorly constructed well. To complicate matters, there was a known breach of a proximal well wastewater pit, which cannot be eliminated as a source.

Even if contamination from oil and gas activities in the Llewellyn et al.[25] study were established unequivocally, the more interesting issue is whether it came from the reservoir or the upper portions of the well. Leaks from the latter are well understood as a possibility from poor practices. The following is a quote from the Discussion section:

> If HVHF fluids did contaminate the water wells, it would be surprising if such contamination were due to fluids returning upward from deep strata, given that (i) this has never been reported (6), (ii) the time required to travel 2 km up from the Marcellus along natural fractures is likely to be thousands to millions of years (31), and (iii) Fig. 6 shows that the Cl:Br ratios in the drinking waters indicate the absence of salts that would be diagnostic of fluids from the Marcellus shale (e.g., flowback/production waters). The most likely way for HVHF fluids to contaminate the shallow aquifers would therefore be through surface spillage of HVHF fluids before injection or by shallow subsurface leakage during injection.

This opinion brings them in line with the thinking of recent investigators.[26,27]

A federal study was conducted by the National Energy Technology Laboratory to assess the likelihood of gas or fracturing fluid constituents communicating from the reservoir up to freshwater bodies.[28] Four different tracers were injected into the fracturing fluid at different stages of the process. Several old wells, accessing a formation approximately 3000 feet above the fracturing zone, were monitored for these tracers. None were detected. Similarly, the tracers were not found in the freshwater zones near the surface. Microseismic monitoring was conducted for this well and an additional seven wells to map the vertical reach of the fractures. Although the vast majority were a few hundred feet long, a single fracture was observed to extend 1800 feet, which was considered anomalous and probably was assisted by an existing natural fracture, a possibility that highlights the need to avoid areas with natural fractures. This observation is also generally supportive of microseismic studies reported elsewhere in this book, in which maximum reaches of approximately 1500 feet were noted.

POSSIBLE CONTAMINATION FROM HANDLING AND STORAGE

Some of the investigative focus has shifted to studies of the potential for contamination from wastewater storage and the handling of neat chemicals involved in rig operations. Wastewater must be stored for periods even where it is to be reused. Unlined pits, such as those at Pavillion, are increasingly in the rear view mirror. However, there are differences concerning single versus double liners, the thickness and quality of the liners, leak detection devices, the nature of the terrain in which the pit is placed, and other factors, many of which are covered in recent regulations.[20] The years prior to proper regulation, especially in Pennsylvania, were strewn with poor wastewater disposal practices. One of the most damaging was dispatch to publicly owned treatment works (POTWs) or municipal water treatment plants. These facilities are not designed to handle high salinity, and the discharge, usually into surface water, invariably had high concentrations. Furthermore, beneficial bacteria were probably impaired by the salinity. Investigations of the downstream effects of POTW discharges into rivers have found evidence of compounds with oil and gas provenance.[29]

A currently favored option is above-ground tanks. These carry between 500,000 and 2 million gallons each. They are portable and assembled on location. The advantage over pits is that postoperational remediation is simpler. For pits, the regulations spell out removal of the liners and associated sludge and disposal of both, and in most cases the land must be restored to its original state. Portable tanks, in contrast, are simply removed at the completion of the mission. Most are double-lined, with sensing between for breach of the first liner. The principal objection by opponents is the risk of catastrophic damage and associated spill. Apache Corporation, a prominent independent oil and gas company, reported the practice of constructing an earthen berm completely surrounding the modular tank, with the intent of containing the fluid in the event of a breach (G. E. King, personal communication, 2011). The berm is removed upon cessation of operations.

In recognition of the likely greater role of handling and storage in groundwater contamination, recent investigations have included these factors in the methods employed. One of the most comprehensive regional studies focused on organic compounds and their possible provenance, as well as the usual inorganic chemicals and geochemical fingerprinting.[27] Fig. 6.3 shows the lateral extent of the studied area and the proximity, or not, to active shale gas wells. All 64 water wells sampled were at private residences and up to 213 meters deep. The organic

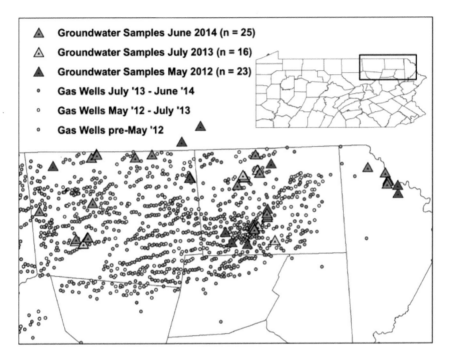

Figure 6.3 Shallow groundwater sample locations and existing active shale gas wells in the Drollette et al.[27] study of northeastern Pennsylvania. Five samples were collected in December 2014 and included in the June 2014 data points. Shale gas well locations were obtained from the Pennsylvania Spatial Data Access. Courtesy: National Academy of Sciences.

compounds were classified into two categories. Gasoline-related organics (GROs) were defined as eluting between 2-methylpentane and 1,2,4-tri-methylbenzene, approximately between nC_6 and nC_{10}, a classic gasoline range, but not necessarily gasoline-derived. Diesel-related organics (DROs) were defined as eluting between nC_{10} and nC_{28}. Results were analyzed with respect to the spatial distribution of the groundwater locations with elevated GROs or DROs with active gas wells and reported surface spills. Low levels of GROs (0 to 8 ppb) and relatively low levels of DROs (0 to 157 ppb) were found. However, ancillary measurements led to the conclusion that DRO contamination was anthropogenic and correlated with distance from gas wells and/or reported surface spills. Direct communication of deep formation water and injected fracturing fluids was ruled out. The cause was probably surface spills of fracturing fluid, the neat chemicals involved in their formulation, or other chemicals used at well sites.

CONCLUSIONS

The evidence to date supports the following conclusions:

- Vertical communication from the fractured intervals to surface fresh water is unlikely when sound practices are followed, especially in light of the typical vertical distance between the two. This applies to liquids as well as gases. Hydraulic fracturing-like chemicals are uncommon in potable water, and where found, they seem to be traceable to surface spills or perhaps shallow underground operations. Stray gas is also uncommon in shallow aquifers, and in a significant portion of these cases the gas has been traced to intermediate formations rather than the deep fractured interval. However, one must guard against operating in areas with vertical faults and fissures. These are detectable by conventional seismic mapping. Such conduits of fluid loss result in lost income. Consequently, the operators have economic if not environmental reasons to comply.
- Analytical methods are in place to precisely investigate concerns about drinking water contamination. The most powerful appear to be the isotopic signatures. The ratio of C^{13} to C^{12} ($\delta^{13}C$) in methane and ethane distinguishes biogenic from thermogenic origin. In some geologic settings, carbon isotopes can correlate with the age of the rock from which the fluid emanated. Similarly, $\delta^{18}O$ plotted against δ^2H in water has been useful for testing whether a mixing relationship exists between deep and shallow fluids. Species characteristics of flowback water are Ba, Br, and Sr. In the Marcellus, the Sr^{87}:Sr^{86} ratio is a clear fingerprint.[30] In general, there is now sufficient research to identify the analytes of importance in each region, and these should be specified in regulations at least as sentinel species.
- Poorly constructed, aged, or inadequately abandoned wells can leak fluids. Wells constructed to the latest regulatory specifications are unlikely to leak. Gas leakage is more likely than liquid leakage. Studies that included baseline testing have not detected leaked fluids or even methane in freshwater bodies. Retrospective studies in the Marcellus have not detected liquid contamination from fracturing operations into shallow potable water.[27] However, continued vigilance is important to identify any future breakthrough from recent or ongoing fracturing operations.

- Baseline testing of water sources proximal to intended drilling activity must be made mandatory. The analytes should be standardized across regions. Each region may have sentinel species identified and could use these for follow-up testing over time with the intent to conduct a full suite in the event a sentinel shows positive.
- Increased emphasis is needed on detecting and preventing spills from surface handling of neat and mixed fluids. Detection of VOCs must receive emphasis. Chapter 4 described a new approach for cost-effective deployment of sensitive VOC detectors.
- Flowback and produced water is most effectively handled through reuse for the same operation. Eventual disposal can be in UIC Class II wells, where available, provided that feasible measures are taken to prevent induced earthquakes. Where injection disposal is not feasible, the technology exists to clean the wastewater for surface discharge or other use.

REFERENCES

1. King GE. *Thirty years of shale gas fracturing: what have we learned? SPE 133456-MS*. The Woodlands, TX: Society of Petroleum Engineers; 2010.<www.onepetro.org> [accessed 14.03.12].

2. Myers T. Potential contaminant pathways from hydraulically fractured shale to aquifers. *Ground Water* 2012;**50**(6):872–82.

3. Engelder T, Cathles LM, Bryndzia LT. The fate of residual treatment water in gas shale. *J Unconvention Oil Gas Resour* 2014;**7**:33–48.

4. Gregory KB, Vidic RD, Dzomba DA. Water management challenges associated with the production of shale gas by hydraulic fracturing. *Elements* 2011;**7**(3):181–6.

5. King GE. *Hydraulic fracturing 101: what every representative, environmentalist, regulator, reporter, investor, university researcher, neighbor and engineer should know about estimating frac risk and improving frac performance in unconventional gas and oil wells. SPE 152596*. The Woodlands, TX: Society of Petroleum Engineers; 2012.<www.onepetro.org> [accessed 14.03.12].

6. LeBas RA, Shahan TW, Lord P, Luna D. *Development and Use of High-TDS Recycled Produced Water for Crosslinked-Gel-Based Hydraulic Fracturing. SPE 163824*. The Woodlands, TX: Society of Petroleum Engineers; 2013.<www.onepetro.org> [accessed 22.05.16].

7. FracFocus. *Chemical disclosure registry*. <http://www.fracfocus.org>; 2013 [accessed 20.06.16].

8. Halliburton. *Fluids disclosure*. <http://www.halliburton.com/public/projects/pubsdata/Hydraulic_Fracturing/fluids_disclosure.html>; 2016 [accessed 24.04.16].

9. Coleman NP. Produced formation water sample results from shale plays. <https://www.epa.gov/sites/production/files/documents/producedformationwatersampleresultsfromshaleplays.pdf>; 2012 [accessed 24.04.16].

10. Horsak RD. The use of electro-coagulation technology to treat hydrofracturing flowback water and other oil and gas field wastewaters. <http://www.marcellusshalewatergroup.com/wp-content/uploads/2014/05/World-Water-Day.pdf>; 2014 [accessed 22.5.16].

11. Ellsworth WL. Injection-induced earthquakes. *Science* 2013;**341**(6142):1225942.

12. Walsh III RF, Zoback MD. Oklahoma's recent earthquakes and saltwater disposal. *Sci Adv* 2015;**1**(5):e1500195.

13. Maloney KO, Yoxtheimer DA. Production and disposal of waste materials from gas and oil extraction from the Marcellus Shale play in Pennsylvania. *Environ Pract.* 2012;**14**:278–87.

14. McCutcheon JR, McGinnis RL, Elimelech M. A novel ammonia-carbon dioxide forward (direct) osmosis desalination process. *Desalination* 2005;**174**:1–11.

15. Winter D, Koschikowski J, Wieghaus M. Desalination using membrane distillation: Experimental studies on full scale spiral wound modules. *J Membr Sci* 2011;**375**:104–12.

16. DiGiulio DC, Wilkin RT, Miller C, Oberley G. Investigation of ground water contamination near Pavillion, Wyoming [Draft Report, U.S. Environmental Protection Agency, Office of Research and Development, National Risk Management Research Laboratory, Ada, OK and Region 8, Denver CO]. <http://www2.epa.gov/region8/draft-investigation-ground-water-contamination-nearpavillion-wyoming>; 2011 [accessed 24.04.16].

17. Wyoming Oil and Gas Conservation Commission. *Pavillion field well integrity review.* <http://wogcc.state.wy.us/pavillionworkinggrp/PAVILLION_REPORT_1082014_Final_Report.pdf>; 2014 [accessed 22.05.16].

18. Wyoming Department of Environmental Quality. Pavillion, Wyoming *domestic water wells draft final report and palatability study*. [Prepared by Acton Michelson Environmental, Inc., El Dorado Hills, CA.] <http://deq.wyoming.gov/wqd/pavillion-investigation/>; 2015 [accessed 01.03.16].

19. DiGiulio DC, Jackson RB. Impact to underground sources of drinking water and domestic wells from production well stimulation and completion practices in the Pavillion, Wyoming, Field. *Environ Sci Technol* 2016;**50**:4524–36.

20. North Carolina Regulations. *Sections 15A NCAC 05H*. 1611 (well installation, including testing cement integrity), 15A NCAC 05H. 1609 (surface casing standards) and 15A NCAC 05H. 1504 (pits and tanks). <http://reports.oah.state.nc.us/ncac/title%2015a%20-%20environmental%20quality/chapter%2005%20-%20mining%20-%20mineral%20resources/subchapter%20h/15a%20ncac%2005h%20.1609.html>; 2015 [accessed 22.05.16].

21. Osborn SG, Vengosh A, Warner NR, Jackson RB. Methane contamination of drinking water accompanying gas-well drilling and hydraulic fracturing. *Proc Natl Acad Sci* 2011;**108**:8172–6.

22. Warner NR, Jackson RB, Darrah TH, Osborn SG, Down A, Zhao K, et al. Geochemical evidence for possible natural migration of Marcellus formation brine to shallow aquifers in Pennsylvania. *Proc Natl Acad Sci* 2012;**109**(30):11961–6.

23. Kendall C, Coplen T. Distribution of oxygen-18 and deuterium in river waters across the United States. *Hydrol Process* 2001;**15**:1363–93.

24. Dresel P, Rose A. *Chemistry and origin of oil and gas well brines in Western Pennsylvania*. Pennsylvania geological survey. 4th series Open-File Report OFOG 10-01.0. 2010:48; 2010.

25. Llewellyn GT, Dorman F, Westland JL, Yoxtheimer D, Grieve P, Sowers T, et al. Evaluating a groundwater supply contamination incident attributed to Marcellus Shale gas development. *Proc Natl Acad Sci* 2015;**112**(20):6325–30.

26. Darrah TH, Vengosh A, Jackson RB, Warner NR, Poreda RJ. Noble gases identify the mechanisms of fugitive gas contamination in drinking-water wells overlying the Marcellus and Barnett Shales. *Proc Natl Acad Sci* 2014;**111**:14076–81.

27. Drollette BD, Hoelzer K, Warner NR, Darrah TH, Karatum Osman, O'Connor MP, et al. Elevated levels of diesel range organic compounds in groundwater near Marcellus gas operations are derived from surface activities. *Proc Natl Acad Sci* 2015;**112**(43):13184–9.

28. Begos K. *DOE study: Fracking chemicals didn't taint water*. Associated Press, July 19, 2013. <http://bigstory.ap.org/article/ap-study-finds-fracking-chemicals-didnt-spread> [accessed 28.05.16]. [This is a news story on the NETL preliminary report.]

29. Warner NR, Christie CA, Jackson RB, Vengosh A. Impacts of shale gas wastewater disposal on water quality in Western Pennsylvania. *Environ Sci Technol* 2013;**47**:11849–57.

30. Vidic RD, Brantley SL, Vandenbossche JM, Yoxtheimer D, Abad JD. Impact of shale gas development on regional water quality. *Science* 2013;**340**(6134):1235009.

PART *III*

Improving Economics of Recovery

There are no facts, only interpretations.
   (Es gibt keine Fakten, nur Interpretationen.)
                              —Friedrich Nietzsche, Notebooks 1886—1887

CHAPTER 7

Illuminating the Reservoir

In estimating the size of hydrocarbon reservoirs, Nietzsche was spot on. Resource estimates are quoted in probabilities. For example, the so-named P50 value is that with a confidence factor of 50%. Similarly, there will be P90 values and so on. The more light we shed on the reservoir, the better the interpretation, in part because the confidence is greater. In most hydrocarbon plays, the resource estimates tend to rise with the conduct of development because new data are produced.

First, some definitions. A *resource estimate* is the quantity of hydrocarbon likely to be in place, whether economically recoverable or not. A *reserves estimate* is the part that is economically recoverable using the current technology. Reserves tend to grow with better data and newer technology. Reserves are the most important determinant of the value of an oil and gas company. For this reason, companies retain expert-level competency in this aspect of the enterprise. Larger companies even produce their own analytical tools. Most other operations are farmed out to service companies. Even the data acquisition required for the reserves computation is outsourced. By and large, the interpretive portion is retained in-house even though the software may be developed by the service companies.

The reserves estimation is classified into two parts: the quantity of recoverable fluid present and the facility with which it can be made to flow into the wellbore. In this chapter, we concentrate primarily on the first part, which is often broadly classified as formation evaluation. However, in the discussion of unconventional reservoirs, we discuss the second part because the permeability has to be artificially created by hydraulic fracturing.

FORMATION EVALUATION OF CONVENTIONAL RESERVOIRS

We limit the discussion to conventional reservoirs and later identify the differences in unconventional reservoirs.

Sustainable Shale Oil and Gas. DOI: http://dx.doi.org/10.1016/B978-0-12-810389-0.00007-3

The Four Steps
Identify Potential Hydrocarbon-Bearing Portions

In conventional reservoirs, producible hydrocarbons are found exclusively in sandstones and carbonates because these have the necessary porosity to hold economically relevant proportions. Rarely are there pure formations of these minerals. Intercalation with clay-like minerals is common. This can confuse the evaluation. In any case, identification is initially based on the natural radioactivity form species in the rocks. The targets are radioisotopes with sufficiently long half-lives and flux abundant enough to be detected economically. In our application, they are ^{40}K with a half-life of 1.3×10^9 years, ^{232}Th with a half-life of 1.4×10^{10} years, and ^{238}U with a half-life of 4.4×10^9 years. These are all gamma emitters, and the measurements are typically from the energy range 0.5 to 3.0 MeV (million electron volts). The simplest measurement is the total from all sources, including some from Compton scattering in the matrix.

The measurement is a lithology indicator because these radioactive species are found almost exclusively in clay, which itself comprises an aluminosilicate of Na and K. Accordingly, substantial absence of natural gamma radiation is an indication of possible reservoir rock. A confusing factor is that the silica or carbonate reservoir rock often has intercalations of clay, which enhance the count. This is in part why gamma traces, known as a log, are rarely examined in isolation from other logs. The measurements are commonly made using scintillation detectors. These comprise doped NaI (sodium iodide) crystals, which produce photons with incident gamma radiation. The photons are counted. In a more sophisticated version of the tool, the energy of incoming radiation can be measured to assess the relative proportions of radioactive species. Curiously, a much older technology, Geiger–Muller tubes, is sometimes used in measurements made while drilling (MWD) because of the inherently robust nature of the device. These tubes are the basis for the handheld counters used by uranium prospectors of yore.

Estimate the Porosity of the Rock

The relative volume of open space in rock is the porosity, and it is expressed as a percentage. In a reservoir, this space will be filled with water or hydrocarbon or both. Porosity is estimated by using two different measurements: neutron and density. Neutron porosity involves bombarding the rock with high-energy neutrons ($\sim 10^6$ eV). These may be produced from a chemical source such as americium—beryllium (AmBe). The source strength is in the vicinity of 1 Curie in MWD and several times that if made on a wireline. The neutrons may also be produced in a neutron generator downhole. In either case, when the high-energy neutrons collide with like-size molecules of hydrogen, energy transfer occurs. (Think billiard cue ball transferring momentum to a numbered ball; the cue ball almost stops). The incident neutrons are reduced in energy (to ~ 0.025 eV). These "thermal" neutrons are detected on the same tool that emitted the original radiation. The tools are calibrated to estimate the total hydrogen ion concentration in the rock. However, one is still left in doubt as to whether the hydrogen is associated with water or hydrocarbon. To decipher this, a resistivity measurement is made.

Estimate the Resistivity of the Rock

Water in reservoir rock is almost always very saline. Consequently, the resistivity is low. By contrast, hydrocarbons are very resistive. This essential contrast is used to determine what proportion of rock is filled with hydrocarbon. Expressed as a percentage, it is referred to as the *hydrocarbon saturation*. In principle, a higher value is preferred, but there can be complications. One complication occurs when sand and shale are intercalated finely, thus confounding the measurement. There is also the interesting case of a formation comprising turbidites. This is likely the most common type of reservoir rock offshore. Prior to approximately 1986, many of these prolific formations were passed over because of the absence of proper measurement techniques. The following box describes their story and a lesson highlighting the role of new analytical techniques in unleashing commercial endeavors.

Unlocking Turbidites

In 1985, MWD was still in a fledgling state. Natural gamma measurements and directional sensing were both commercial. However, the all-important resistivity measurement was not considered quantitative enough to displace wireline logs. Then a new investigative tool was introduced by NL Industries, the electromagnetic wave resistivity (EWR) sensor. It launched a 2-MHz wave off a portion of the drill string, and the phase changes in the return wave were proportional to the resistivity of the formation. A feature of the sensor was vertical resolution of a mere

12 inches. At that time, wireline sensing had a short reading sensor known as short normal, which was not quantitative. The quantitative sensor, the deep induction tool, had vertical resolution of a few feet and was the primary measurement to calculate hydrocarbon saturation.

An early test of the EWR sensor was on the Cougar Platform of Shell Oil. Of the three prospective sands, only the A and C zone sands were being produced. The B zone registered on the wireline log with very low resistivity and therefore was presumed to be noncommercial. When the EWR log was run routinely while drilling through the B zone on the way to the C zone, a surprising feature appeared. The log showed a uniform sequence of relatively high and very low resistivity zones, each a couple of feet thick. Clearly, sand and shale were in sequence, and the deep induction tool was averaging many layers. In actuality, the EWR was identifying the B sand as potentially prospective. There was still some doubt because the resistivity was still low enough to compute a hydrocarbon saturation of only approximately 30%. However, when put on production, it delivered oil with almost no water cut. It seems the water was bound to the rock and only oil was mobile, and it was very prolific. No more wireline logs were run on that platform, only MWD.[1] This was an industry first and created quite a stir. A new measurement technique had unleashed a class of reservoirs known as turbidites, which are ubiquitous in waters off the continental shelf. This success also launched MWD into the mainstream of formation evaluation tools, largely replacing wireline logs in development wells.

Estimate the Density Porosity of the Rock

The resistivity measurement estimates the relative proportion of hydrocarbon and water. One then needs to know how much of the rock is fluid filled. This is achieved by measuring the bulk density of the formation. The sensor comprises a source of medium-energy gamma rays, most commonly from a ^{137}Cs chemical source. The gamma rays collide with electrons in the rock elements and are scattered by the Compton effect. The number of scattered gamma rays reaching the detector is directly proportional to the electron density of the material, which in turn is proportional to the bulk density. The detector is placed some distance away from the source in the same tool that launched the gamma rays. If the matrix density is known precisely, then the porosity can be computed by

$$\phi = \frac{\rho_{ma} - \rho_b}{\rho_{ma} - \rho_f}$$

where ma is the rock matrix; b is the bulk density measured by the sensor; and f is the fluid in the pores, usually salty water, during calibration. The bulk density of common matrix rocks is 2.65 g/cm^3 for sandstone, 2.71 g/cm^3 for limestone, and 2.87 g/cm^3 for dolomite.

Log Evaluation

Porosity measurement is difficult when the fluid is gas. The neutron porosity is calibrated with water-filled pores. Gas molecules are less dense than oil or water. Thus, the neutron tool will measure a low porosity when gas filled. On the other hand, the density porosity will read high. In the nearby water or oil leg, they should read essentially the same. Consequently, one way of distinguishing between oil and gas is this "crossover" of the two traces in the case of gas.

All of the four measurements described previously are plotted side by side in a vertical mode, with formation depth on the vertical axis. This plot is known as a log. Fig. 7.1 is a schematic example without some features found in actual logs. Qualitative interpretation is done first, using the gamma to identify the likely reservoir rock. In Fig. 7.1, this is the left trace in green. At approximately 4185 feet, the gamma drops to the level associated with sandstone. The deep reading resistivity shows as high from approximately 4185 to approximately 4195 feet. This indicates some sort of hydrocarbon. Shifting to the third trace, neutron porosity is shown in blue and density porosity in red. The traces cross over in the portion shown by the orange shading. This is a region with gas, and the relatively high resistivity supports this finding. At approximately 4191 feet, the porosity traces come together and the resistivity remains high. Thus, this is an oil zone. At approximately 4195 feet, the resistivity drops but the gamma stays low. Thus, this is sandstone with water in it. The porosity traces come together in the clean sandstone after approximately 4191 feet as they should.

In conventional reservoirs, this sort of sequence is quite common: gas on top, oil in the middle, and water below. The gas will often be produced last, leaving it to exert natural pressure to drive up the oil. Similarly, every effort will be made to not produce the water portion. One can see, therefore, that even this simple log is adequate for deciding where to produce. This simple combination is known in the industry as the triple combo log.

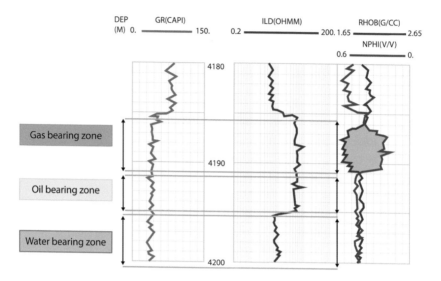

Figure 7.1 Schematic trace of a triple combo log.

FORMATION EVALUATION OF UNCONVENTIONAL RESERVOIRS

Unconventional reservoirs include shale, tight sands, and coal bed methane. In this book, we ignore the last one, except for minor allusions in context, and discuss the first two, with a focus on shale oil and gas because these are the most relevant to today's situation. Also, coal bed methane brings a different set of environmental issues, and we limit the scope to tight reservoirs.

What makes shale reservoirs different is that they comprise the source of all hydrocarbons, and only recently have they emerged as a direct resource. All conventional hydrocarbons are found in porous sand and carbonate. They migrated there over time from the source rock and were trapped in place by some physical sealing mechanism, usually an impermeable layer of shale.[2] The source rock was where the organic matter collected, and over millions of years of high pressure due to progressive soil deposition above, it was converted first to kerogen. This early immature hydrocarbon is occasionally harvested in the form of *oil shale* (as distinct from *shale oil*). Over time, the kerogen breaks down to oil at first, and then finally the molecules get smaller all the way down to methane. The simplest way to think of this is to understand that the vast majority of oil has the following formula, at least to begin with:

$$C_nH_{2n+2}$$

where n is 1 for methane (giving CH_4), 3 for propane (C_3H_8), and so on.

The continuum proceeds to higher numbers. Oil is defined as a mixture of molecules with n in excess of approximately 25. Accordingly, kerogen first breaks down to larger oil molecules and then finally to methane. However, because this is a continual process, shale oil always has some associated smaller molecules. This is often flared today, which is discussed in Chapter 2. Similarly, shale gas usually has some larger molecules known as natural gas liquids. When it does not, it is called dry gas, which represents the most thermally mature state of the process.

The problem with reservoir evaluation begins with the fact that this rock more resembles the rock that seals the reservoir fluids in place than a reservoir. Conventional reservoirs have permeability down to 1 millidarcy (mD), and the really good ones are single-digit darcy. (The darcy is the most common unit of permeability in the industry.) High-quality shale reservoirs are typically at 0.001 mD; in other words, the best shale reservoir is 1000 times worse in permeability than the worst conventional reservoirs. In the main, one needs to induce permeability to produce from such rock. This is accomplished with hydraulic fracturing, without which production would be well nigh impossible in most cases. Natural fracture systems can help as well, and the Austin Chalk is a case in which they are sufficient in themselves (this is a tight rock but not shale). The reliance on hydraulic fracturing brings into focus properties of rocks that are more amenable to this process. Ordinarily, one is not too concerned about the properties of shale, but in this case the details matter.

Shale is some mixture of at least three minerals: quartz (SiO_2) or feldspar (silicate with K, Al, and Na); a carbonate of some sort, usually calcite ($CaCO_3$) or dolomite ($CaCO_3.MgCO_3$); and clay (hydrous aluminosilicates with Na and K). The proportions of these determine the suitability for fracturing operations. Fig. 7.2 shows the compositions for some well-known deposits of shale. The ternary diagram is

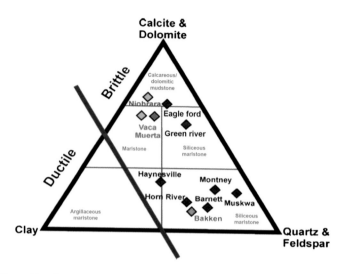

Figure 7.2 Composition of various producing shale deposits. Courtesy: Colorado School of Mines.

read as follows: The apex corresponds to 100% of that mineral, and the center of the triangle represents precisely one-third of each mineral. Note that compositions anywhere in the vicinity of the clay apex (left of the red line) are not viable because they are too ductile. Similarly, but less critically, the apex of each of the other two is also out of bounds as being too brittle. One of the most prolific deposits, the Eagle Ford in Texas, is in the carbonate sector. Interestingly, so is the Vaca Muerta in the Neuquén province of Argentina, which in the view of one of the authors (VR) is more prospective than any single play in the United States.

Brittleness

In the past, the focus was on shales with mechanical properties at the other end of the spectrum: ductile shales. These are present in the wellbore and can be an impediment to drilling because they tend to swell and slough off. The swelling is occasioned by permeation of water from the drilling fluids into the clay. Oil base muds were developed to combat this. They act by forming a semipermeable membrane effect so that water flows only in one direction—into the wellbore.[3] The objective was to preserve the mechanical integrity of the shale layers, which is quite distinct from the intent in fracturing.

Because the study of brittleness of shale is relatively new, there is disagreement regarding the contributing factors. All agree that quartz contributes to brittleness and that clay is the primary constituent imparting ductility. Calcite is relatively ductile, but it is the dolomite that engenders some debate. In our view, it should be considered brittle. A phenomenological basis for that opinion is the high quality of the reservoirs in that part of the ternary plot. Partly because of these differences in opinion, mineralogy is seen as indicative but not deterministic of fracturing potential. For that, direct mechanical measurements are preferred, usually on core, as discussed later.

In other materials, the right degree of brittleness is assessed by using the concept of fracture toughness. In metals and alloys, this is determined through the Charpy impact test. For several reasons, the oil and gas industry would be well advised to adopt the term *toughness* rather than brittleness. Fracturing has two steps, initiation and propagation. The mechanical properties of the rock are different for the two mechanisms. Some plasticity may actually be of value in the initiation stage. Also, natural fractures could play a role in initiation. Some authorities are proposing the term *fracability*, but standard definitions are still lacking. A commonly used parameter is the brittleness index BI:

$$BI = e_{el}/e_{total}$$

where e_{el} is the elastic strain, and e_{total} is strain at failure.

Coring and Core Analysis

Although core analysis is the most sound method to obtain the necessary data, it is expensive. Consequently, industry efforts are focusing on wireline and seismic techniques. These primarily center on estimation of rock's elastic properties by using sound waves, both P (longitudinal) and S (shear) waves.[4] A core is a cylindrical sample of rock acquired in a relatively undisturbed state. The most useful sort is known as whole core. This is acquired by attaching a core barrel to the drilling string, with a special bit on the end. This bit cuts the rock in such a way as to push rock up the core barrel. The barrel is retrieved, and tests are performed on the core. In recent advances, some of these tests are done while the rock is still in the barrel in a preserved state, but usually the cylinder of rock is removed and analyzed. A small plug sample is drilled out at intervals for special core analysis.

Another type is sidewall core. In this case, a tool is deployed in the open borehole (i.e., the well shaft), and a rotary drill carves a cylindrical plug out of the wall. The locations are determined by wireline or MWD logs run earlier. Much cheaper than whole core, this is preferred provided the quality of information is sufficient. In both cases, core examination is often used to calibrate the logs.

Cores acquired either way are examined for much the same parameters as logging. Whereas logging measurements are largely inferential, these are direct. Porosity is directly measured by injecting a known fluid into the empty pores. Permeability is measured by actual fluid flow through the core sample (usually on a plug). Relative permeability can be measured with ease. In the case of mixed fluids, it might be important to know which fluid is more mobile in the matrix. Rock's mechanical properties are also acquired on cores. Petrophysicists often refer to these measurements as ground truth. Fluids under pressure in the reservoir are unlikely to be in the same state when the core is recovered at the surface. This is compensated for by modeling. Special core barrels that keep the core under pressure have been devised, but high cost keeps their use to a minimum. One company employed a sponge core system wherein the core barrel was lined with an absorbent material to capture the fluids. Although cheaper than pressure coring, utilization is still low.

Analyzing Unconventional Reservoir Core

The importance of evaluating core is that log data, whether on wireline or while drilling, require calibration with core much more than in conventional reservoirs. As previously discussed, three critical evaluations are needed: estimate of hydrocarbons in place, rock properties to determine suitability for fracturing, and reservoir quality for a host of factors determining economic recovery of a high fraction of the oil or gas in place.

The parameters being measured are for the most part the same as those for conventional reservoirs.[5] However, the techniques required have to be refined. The permeability is at least four orders of magnitude less than in normal reservoirs. Furthermore, porosity is by nature variable. Loucks et al. described this in an important work.[6] Almost all researchers use their own classification of porosity. Loucks et al. postulate three types: pores associated with organic matter, pores between the grains (interparticle), and pores within the grain (intraparticle), with the latter two being unrelated to organic matter. One can see that the gross porosity derived from traditional logging tools could be misleading. In fact, when newer services are offered that estimate just the organic matter porosity (Phi_{OM}), the operator will no doubt want to validate the results against those from the conventional log. The discrepancy will need explanation. The new offering will face the same hurdles that the EWR faced in the early days (see the previous text box) when the values were compared to the only known quantitative tool at the time, the deep induction resistivity. One of the authors (VR) lived through the teething pains of that experience.

One feature in favor of the new offerings estimating Phi_{OM} is that the microscopy technique also allows evaluation of the nonproductive intergranular porosity. Furthermore, some believe that these pore spaces hold water and are responsible for the water cut with production. Their identification, aside from making the total porosity more comparable to conventional measurements, offers the possibility of avoiding those portions of strata to reduce water cut. These different porosities can be distinguished only with high-resolution microscopy. Enabling this was the development of argon-milling surface preparation and the use of the field-emission scanning electron microscope (SEM). Now pores on the scale of single-digit nanometers can be seen.[7] This scale was necessary for the deciphering of the different types of porosity. SEM had been in use for decades, but the associated analytics on massive quantities of data are now feasible for a reasonable cost.

Digital Rock Physics

This field has seen rapid growth since the shale oil and gas explosion. However, the early development and applications of the technology preceded the shale revolution by at least a decade.[8-11] The essential elements of digital rock physics (DRP) are high-resolution two- and three-dimensional (2-D and 3-D) imaging; sophisticated modeling and associated data analytics; and relatively low-cost, high-speed computing and graphics techniques originally developed for gaming. The key concept is "image and compute," which means that the physics of fluid flow, electrical conductivity, elastic properties, etc. has to be understood and that numerical solvers must be available to allow the complex pore space and solid mineral framework of real rocks to be accommodated. Because of the complexity of rocks in general and shales in particular, the imaging often must be done at multiple scales to capture all the important variability and heterogeneity.

Although DRP makes use of modern commercial imaging tools such as the SEM, focused ion beam (FIB)–SEM, and 3-D computed

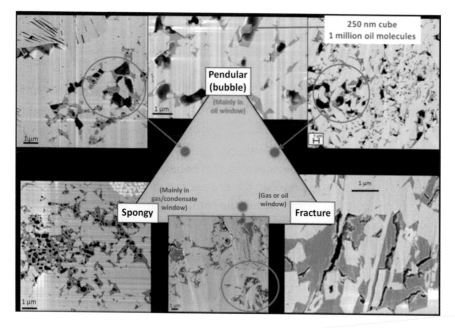

Figure 7.3 Types of organic matter porosity. Courtesy: Ingrain Inc.

tomography (CT) imaging, due to the particular demands of the "compute" requirement of DRP, and the complexity of real rocks, special and unique methods of sample preparation and imaging have been developed.

DRP techniques have allowed description of detail not possible in the past. Fig. 7.3 shows a classification of organic matter porosity first reported by Walls and Sinclair.[12] Whereas Loucks et al. classified porosity into organic matter, intergranular, and intragranular, Walls and Sinclair further break down the organic matter part into three areas:

- Relatively compact, almost solid matter often has fractures inside the organic material and so is likely productive. In Fig. 7.3, this is shown in the bottom right corner labeled "fracture."
- Pendular porosity, almost like bubbles, has the organic matter substantially filling the smaller intergranular spaces, whereas the larger pores remain open and unfilled. This is shown in the top corner of Fig. 7.3. This type of porosity is usually associated with oil as the fluid.
- Spongy porosity, in the bottom left corner of Fig. 7.3, is usually associated with the most thermally mature state of the original kerogen, which is gas, as discussed previously.

Although there appears to be a trend of increasing thermal maturity in moving from the pendular organic porosity to more spongy textures, it is not clear why some shales exhibit more or less of the fracture-style organic porosity. Investigators have observed that all three types of porosity may occur in any given sample.[13] It may be related to the type of organic matter, in which some formations have "stiffer" kerogen that does not form bubble-type pores but, rather, tends to fracture. Note that kerogen or bitumen can form thin cracks from devolatilization at ambient conditions. These are usually narrow cracks with smooth sides and sharp end points, and they are not believed to have been formed in situ. The proportions of the different organic porosity types relative to intergranular porosity will then likely dictate the performance and the type of fluid one would expect to produce.

A striking conclusion from the previous observations is that this technology sheds substantially more light on the reservoir than even enhanced conventional methods could hope to accomplish. For example, the porosity associated with organic matter (PAOM) can be quantified by DRP methods, but there is no comparable conventional core analysis method to do this. PAOM is relevant because it is almost always hydrocarbon filled. Inter- and intragranular porosity, on the other hand, will contain some water, and it may be 100% water filled. In plays such as the Wolfcamp, where water production is an issue, the completion design would usefully target the regions of high PAOM and not just those with high porosities.

Case Study of Digital Rock Physics From the Permian Basin
The case study described here used archived core from the Bureau of Economic Geology in Austin, Texas. The well (Wolfcamp 1) was located in the Midland Basin, Texas. The study of this core was done in four steps.

Dual-Energy X-Ray CT Imaging and Petrophysics
From X-rays produced at different energy levels, continuous whole core scans were calibrated to produce bulk density (RHOB) and photoelectric factor (PEF) values at approximately 0.3-mm resolution. The RHOB and PEF values combined with core spectral gamma ray (SGR) data are used to compute porosity and organic content and to delineate mineralogy.

Fig. 7.4 shows the core scan data and petrophysical interpretation from Wolfcamp 1. The initial scan can be done within 2 days of acquisition, while the core is still in the aluminum core barrel. The next steps of imaging and associated interpretation take approximately 3 days to complete. These timescales are sufficient for useful operational decisions, particularly if batch drilling is conducted.

The RHOB, PEF, and SGR values are utilized to determine the locations for plug sampling. Before the availability of these petrophysical interpretations, which are made while the core is still in the core barrel, plugs were taken at fixed depth intervals of 1 to 3 feet. The vertical heterogeneities in shale reservoirs render the geometric approach particularly unsuitable.

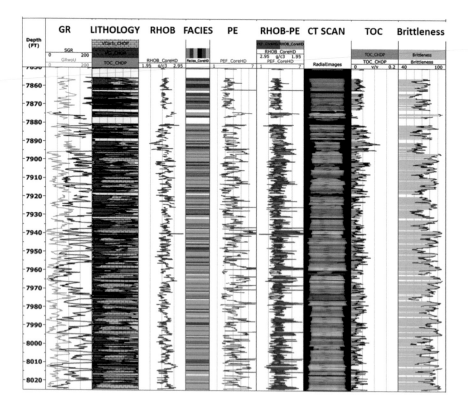

Figure 7.4 Wolfcamp 1: Dual-energy CT plus spectral gamma scan, with tomofacies and petrophysical interpretation. DEN, density; LITH, lithology; PEF, photoelectric factor; SGR, spectral gamma ray; TOC, total organic content. Courtesy: Ingrain Inc.

Analysis of Core Plugs

The plugs are imaged to estimate relevant rock properties such as bulk density, porosity, organic matter content, and clay volume. The imaging comprises a combination of micro-CT scanning and SEM at multiple scales. Fig. 7.5 depicts multiscale imaging of a plug: inch-scale macro-CT on the left, micrometer-scale SEM in the middle, and nanoscale SEM on the right.

High-quality images permit the detailed analysis of rock properties. The analysis includes estimates of inter- and intragranular porosity, organic porosity, solid organic material, and high-density minerals such as pyrite (Fig. 7.6). Other useful interpretations from the images are pore size distribution, pore aspect ratios, and the fraction of solid organic material that has been transformed to porosity.

Rotary Sidewall Samples and Drill Cuttings

DRP processes are used most commonly on whole core but also on sidewall cores. At least one service company is able to preserve fluid pressure and even a sample of fluid in the retrieved sidewall core. This last is a useful attribute for the analytical method discussed in

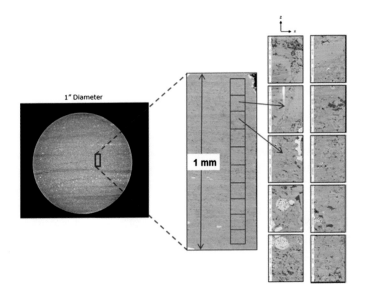

Figure 7.5 Multiscale imaging of a 1-in.-diameter plug sample. Courtesy: Ingrain Inc.

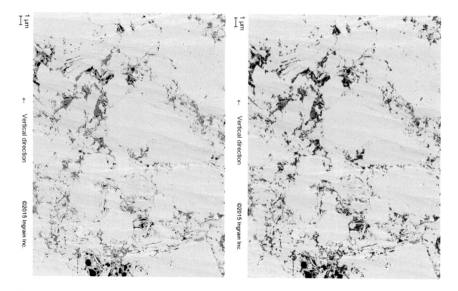

Figure 7.6 (Left) Secondary electron SEM image of Wolfcamp 1 shale showing pores, organic material, and mineral grains. (Right) Analyzed image showing intergranular porosity (red; 2.29%) and PAOM (blue; 3.05%) for an effective porosity of 5.34% (clay-bound water porosity not included in effective porosity). Courtesy: Ingrain Inc. Walls JD, Hintzman T, Guzman B. Shale porosity analysis from SEM imaging and traditional lab methods. AAPG Annual Convention and Exhibition, June, 2016, Calgary, Alberta. <http://www.searchanddiscovery.com/abstracts/html/2016/90259ace/abstracts/2383891.html>; 2016 [accessed 12.07.16].[14]

Chapter 9. However, sidewall coring is less common in horizontal wells because of the cost.

The least operationally intrusive sampling in wells is retrieval of drill cuttings. The most commonly used type of bit is the polycrystalline diamond compact bit. It cuts rock by shearing rather than grinding (as do the other types of bit). Accordingly, the chips of rock are intact and larger than the grain and pore sizes in the reservoir. This enables SEM imaging and DRP methods. DRP analysis requires careful collection of representative rock, sample preparation, and selection of imaging locations. Fig. 7.7 shows the results of the evaluation of drill cuttings in a horizontal wellbore. Each bar in the figure is 30 feet wide. Significant sections of the horizontal show little heterogeneity in TOC, but in these sections there is some variability in brittleness index. In Chapter 9 we demonstrate lateral heterogeneity in a horizontal well using DNA sequencing methods. In the future, methods such as DNA sequencing and DRP will be significant in targeting only the most productive zones for fracturing rather than using the current practice of geometric placement of zones. This has the potential to reduce fracturing costs by up to 30%.

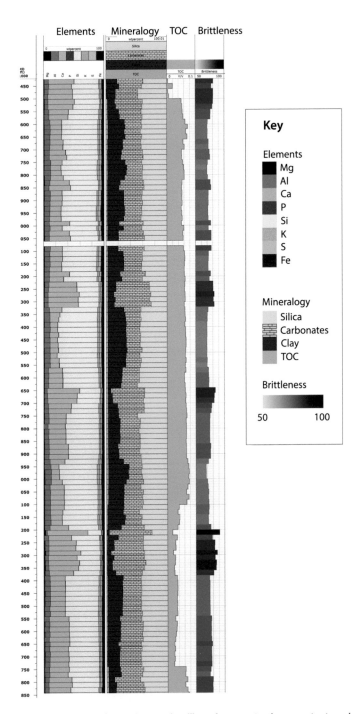

Figure 7.7 Drill cuttings analysis from a horizontal wellbore demonstrating heterogeneity in rock properties.
Courtesy: Ingrain Inc.

Permeability From 3-D FIB–SEM

Nanometer-scale FIB–SEM pore and matrix imaging in 3-D is the first step in estimating permeability. The next is image processing and creation of digital rock volumes. To ensure consistency, the same rock volume is used for all the analyses to derive rock properties. These properties may include the different classes of porosity, pore size, and throat distribution and permeability, both horizontal and vertical.

A classic petrophysical technique is to establish the relationship between porosity and permeability for each of the primary producing formations or facies. This is useful because porosity in rock is more easily measured than permeability. Once the relationship is established, porosity may be used to infer permeability. Fig. 7.8 shows this relationship for the Wolfcamp 1 well. DRP will also distinguish between the shale pore types and identify the ones in the producing facies. Porosity uniquely associated with organic matter is especially critical to good reservoir quality, as discussed earlier in this chapter. The minimum number of samples required to obtain

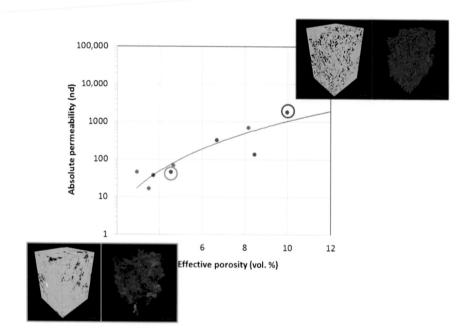

Figure 7.8 Porosity versus permeability data from five FIB–SEM volumes and the computed trend line. Two FIB–SEM volumes are shown for illustration. In the transparency volume (right), the blue color indicates connected porosity. Courtesy: Ingrain Inc.

valid trends depends on the vertical heterogeneity of the target formations and the lateral variability within the play. In Fig. 7.8, the five samples analyzed form a reasonable trend to transform the computed porosity to a permeability score.

Upscaling From SEM to Whole Core

The final step in this case study was to compute continuous porosity and permeability to be displayed together with the other petrophysical computations in a standard log format (Fig. 7.9). The result combines data and interpretations from each step, beginning with dual-energy X-ray CT imaging of the whole core, proceeding to plug scale study and then to pore scale analysis, and finally integrating all data and interpretations to produce a comprehensive analysis of the whole core.

The foregoing discussion highlights the necessity for core analysis that goes well beyond what the industry has classified as special core analysis. The necessity is driven by the unique attributes of shale reservoirs, which make interpretation difficult. In Chapter 9, we describe yet another unique tool to illuminate these complex reservoirs.

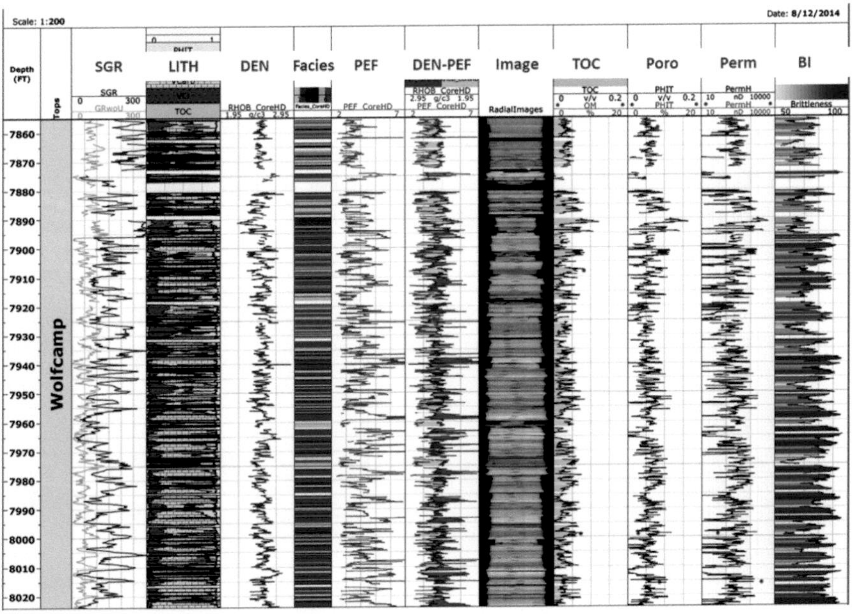

Figure 7.9 Wolfcamp 1 log of porosity, permeability, and brittleness index over cored interval computed from integrated digital rock data. Courtesy: Ingrain Inc.

REFERENCES

1. Greif A, Koopersmith C. Petrophysical evaluation of thinly bedded reservoirs in high angle/ displacement development wells with the NL recorded lithology logging system. The Tenth Formation Evaluation Symposium, Canadian Well Logging Society, October 1985. *The Log Analyst* 1986;**27**(5).

2. Rao V. *Shale oil and gas: the promise and the peril.*. 2nd ed. Research Triangle Park, NC: RTI Press; 2015 [Chapter 1].

3. Hale AH, Mody FK, Salisbury DP. The influence of chemical potential on wellbore stability. SPE-23885-PA. *SPE Drill Completion* 1993;**8**(3).

4. Castagna JP, Batzle ML, Eastwood RL. Relationships between compressional-wave and shear-wave velocities in clastic silicate rocks. *Geophysics* 1985;**50**:571–81.

5. Ubani CE, Adeboye YB, Oriji AB. Advances in coring and core analysis for reservoir formation evaluation. *Pet Coal* 2012;**54**(1):42–51.

6. Loucks RG, Reed RM, Ruppel SC, Hammes U. Preliminary classification of matrix pores in mudrocks. *Gulf Coast Assoc Geol Soc Trans* 2010;**60**:435–41.

7. Loucks RG, Reed RM, Ruppel SC, Jarvie DM. Morphology, genesis, and distribution of nanometer scale pores in siliceous mudstones of the Mississippian Barnett Shale. *J Sediment Res* 2009;**79**:848–61.

8. Yale DP. Network modeling of flow, storage and deformation in porous rocks. *PhD Dissertation*. Stanford University; 1984:167 pp.

9. Dvorkin J, Nur A. Elasticity of high-porosity sandstones: theory for two North Sea data sets. *Geophysics* 1996;**61**(5):1363–70.

10. Arn CH, Knackstedt MA, Pinczewski WV, Garboczi EJ. Computation of linear elastic properties from microtomographic images: methodology and agreement between theory and experiment. *Geophysics* 2002;**67**:1396–405.

11. Keehm Y. Computational rock physics: transport properties in porous media and applications. *PhD Dissertation*. Stanford University; 2003:135pp.

12 Walls JD, Sinclair S. Digital rock physics provided critical insights to characterize Eagle Ford. *Am Oil Gas Rep* February 2011.

13. Walls JD, DeVito J, Diaz E. Digital rock physics. *Oilfield Technol* 2012;**5**(2):25–8.

14. Walls JD, Hintzman T, Guzman B. *Shale porosity analysis from SEM imaging and traditional lab methods*. Calgary, Alberta: AAPG Annual Convention and Exhibition; June, 2016. <http://www.searchanddiscovery.com/abstracts/html/2016/90259ace/abstracts/2383891.html>; 2016 [accessed 12.07.16].

Improving Net Recovery of Fluids

In Chapter 1 we discussed the oil and gas landscape and identified the pressing need for innovation for survival in a low-price environment. Despite the fact that unconventional reservoirs presented unique challenges in exploitation, considerable progress has been made in a short period of time in this slow-to-change industry. Initially, this was in natural gas. In the early 21st century, when the industry started taking off, gas was priced high and domestic demand was not expected to be met without imports. Because natural gas pricing is strictly regional, the expectation of sustained high prices was the stimulus for innovation in producing natural gas from shale. By the end of the first decade, the more difficult target of oil became part of the focus. This was in a pricing environment of oil expected to be more than US$100 per barrel for decades.

Then the wheels came off. The extraordinary success of shale gas exploitation produced a glut that crashed the prices to levels at which dry gas (without material amounts of more lucrative larger molecules) production was largely uneconomical. The emphasis shifted to oil in the second decade of the century. From 2011 to 2014, the United States added 1 million barrels per day (bpd) each year. Against the backdrop of reduced demand in the growth areas of China and India, this caused a crash in oil prices worldwide, abetted by OPEC's decision not to prop up prices by reducing production, as it would ordinarily have done.

It is now clear that the methods employed in the early years are too inefficient in a low commodity price market. We define *net recovery* as the percentage of the fluid in place that is recovered. For both hydrocarbons, the numbers are low compared to production from conventional reservoirs. Oil has the greatest disparity. Net recoveries are averaging 5%, compared to the mid-thirties range for conventional production. This is viewed as the primary target for innovation—to increase the recovery percentage for shale reservoirs initially to the high single digits. In response

Sustainable Shale Oil and Gas. DOI: http://dx.doi.org/10.1016/B978-0-12-810389-0.00008-5

to the price drop, the industry has already made significant gains in operational efficiency. This was briefly discussed in Chapter 1. The breakeven numbers have been reduced, in many cases, to the US$40 to US$55 per barrel range. Further reduction requires improvement in net recovery. This chapter discusses factors that must be considered and methods that can be employed to achieve that operational effficiency.

NATURAL FRACTURES

This rock is different. Porosities are often an order of magnitude less than in conventional reservoirs, and the matrix permeabilities are several orders of magnitude less. The latter is the greater concern. The good news is that virtually all shale reservoirs are naturally fractured.[1] Porosity, permeability, and natural fracturing have complex relationships with production, as discussed later.

Although natural fractures are nearly always present, rarely are they productive in themselves despite having effective permeabilities up to three orders of magnitude greater than the matrix permeability. This is because by and large they are closed, often by mineralization, which acts as a binder. However, the minerals usually do not bond chemically to the fracture faces. Consequently, the faces are prone to open up with the stress of hydraulic fracturing.[2,3] Laboratory experiments have demonstrated that the breakdown pressure is half that in similar rock free of fractures when the bonding mineral is calcite, which is the most common situation.[3] This has been confirmed phenomenologically. The rock mechanics involved are complicated and well beyond the scope of this discussion.[4] Suffice it to say that natural fractures and the interactions with them are critical to success. The most important measurement technique is microseismic monitoring.

Microseismic monitoring is a passive technique that involves using three-axis accelerometers to detect small seismic events. Usually, the devices are placed in an observation well in the middle of the development. The hydraulic fracturing event induces changes in the stress state and pore pressure of the rock around the fracture.[5,6] Shear slippage occurs along the planes of weakness in the rock so affected. This is much the same as in earthquakes, except that it is a micro event. The resulting sound wave is detected by the accelerometers. The locations of these bursts of energy are mapped. In this manner, one can estimate the progress of induced fractures as well as the interaction with natural fractures. This technique was used to establish the fact that fractures from the stimulation process rarely travel greater than 1000 ft in the vertical direction. This finding was key in supporting the general belief that fractures from the reservoir are not conduits for groundwater contamination by fracturing fluid.

Natural fractures have another role in the process. Their presence in the borehole wall is an aid to fracture propagation immediately following initiation by the hydraulic pressure applied.[7] Ordinarily, the pressure needed to initiate a fracture is derived from a rock property known as *fracture gradient*, which is expressed in pounds per square inch per foot (psi/ft). The Barnett prospect has typical true vertical depths ranging from 5000 to 8000 feet. It is also reported to have mostly normal fracture gradients.[8] One would expect fracture initiation pressure to be in the range 2100 to 3400 psi. The presence of natural fractures proximal to the borehole can materially reduce the pressure required to propagate the induced fractures.[3]

An interesting observation is that whole cores from vertical shale wells often do not show natural fractures.[9] This is probably due to the fact that these are separated by several feet when present and are almost always in the vertical direction. A single vertical core could easily miss the fracture. Consequently, whereas coring is the workhorse for determining "ground truth" for petrophysical parameters (see Chapter 7), it appears to be of limited value for this purpose. The Austin Chalk development in the early 1990s was driven by the recognition of the vertical aspect of natural fractures. Vertical wells delivered poor results, but horizontal intersections were very productive. This development was the principal driver of testing and the eventual popularity of horizontal wells. Although that was the economic driver, the key technology enablers were measurement while drilling (MWD), three-dimensional (3-D) seismic brought to the desktop, and steerable systems. All of this happened in a down market much like that of 2015/2016.[10]

Normal 3-D seismic will not resolve the natural fractures seen in shale. However, the progress of induced fractures will be mediated by the natural fractures. This is revealed by microseismic measurements. To the extent that fractures intersect the wellbore, they can be detected with a variety of techniques. In water-base muds, the best technique is imaging by resistivity measurements such as formation microimaging (FMI). The original borehole imaging method was invented by Mobil Oil in the 1980s and used acoustic measurement. Optical measurement with cameras is also available. FMI detects fractures and their orientation. This information can be used to select zones to fracture, but logs with such sophistication are rarely run in shale oil and gas wells for reasons of cost.

HETEROGENEITY IN HORIZONTAL WELLBORES

With very few exceptions, fracturing is done by geometric methods. Fig. 8.1 is a schematic of the operation. Shown is the most common method, which uses a cemented and cased borehole. In principle, if the rock is competent and the borehole is uniform, then the method may be employed in an open hole using a variant mentioned later. This technique involves setting a series of plugs to separate the zones of interest. The plugs are now made of a polymeric material such as polylactic acid and serve to isolate a zone from pressure in a neighboring zone. The procedure involves fracturing the lowest zone, setting a plug above it, and then fracturing the second zone. Each zone is perforated with a perforating gun. This is like bullets being fired to pierce the

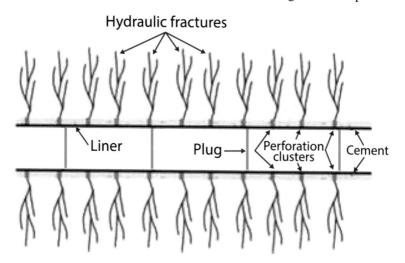

Figure 8.1 The plug and perf method of hydraulic fracturing. Courtesy: World Oil.

casing liner and the cement to expose the formation. (In early embodiments, they *were* bullets; currently, a higher level of sophistication employs shaped charges.) Each zone will have between two and six clusters, which are perforations spaced in a pattern. Then the hydraulic fracturing operation is commenced. The sequence is repeated, resulting in a series of fractured zones. The key is that the zones are equally spaced, as shown in Fig. 8.1. The spacing is usually between 200 and 400 feet. That is, the design is geometric and not based on an understanding of possible heterogeneity in the horizontal. When all the zones have been fractured, the plugs are drilled out in sequence until all the zones are exposed to the production tubing. Some flowback water will come to the surface at each plug drill-out, particularly if the procedure involves a bottom-up circulation to ensure that all the polymer chips are removed.

The alternative to plug and perf uses sliding sleeves to access the formation. The stages are separated by packers, which seal against the formation or cemented casing and prevent communication between stages. The function is similar to that of the plug in the previous technique. In a cased hole, the casing would be perforated at the locations of interest prior to running-in the system. For the first stage, the sleeve is opened at the bottom and the fracture is performed. This is repeated at each stage up the hole. The locations of the stages are predetermined, so the technique is well suited to the geometric approach. However, if a means were found to identify productive zones, this design would be less flexible for targeted zones than the plug and perf method.

Before 2011, adherence to such a brute-force process could well have been understandable. Uneven distribution of natural fractures was known to some extent, but no method existed to identify the distribution economically. Besides, heterogeneity was in the realm of informed conjecture. Then emerged a paper by Miller et al. titled, "Evaluation of Production Log Data from Horizontal Wells Drilled in Organic Shales."[11] They reported on more than 100 shale gas wells in a multiplicity of settings, all horizontal. They demonstrated statistically that there was high variability of production along the wellbore. The measurements were made with a device that measures multiphase flow at each stage. These were all gas wells; thus, in most instances, they also carried some condensate. The data were from all the principal basins—Woodford, Barnett, Fayetteville, Eagle Ford,

Haynesville, and Marcellus—with the highest well count being from the first two (91 of the 112 wells). Key findings included the following:

- Roughly 30% of all clusters had essentially no production.
- Approximately 20% of the clusters produced more than 150% of the average rate.
- The best performers had fewer clusters per stage.
- The optimal stage length was between 300 and 400 feet.

The finding that 30% of all clusters do not produce is often quoted, albeit sometimes erroneously as 30% of all stages.

UNDERSTANDING AND HARNESSING HETEROGENEITY

Knowing that there are heterogeneities is a good first step. However, assessing the source and taking action accordingly have proven more elusive. This is the reason why there are only occasional instances of operators attempting to target zones rather than to fracture geometrically. Heterogeneity may be ascribed to two possible categories: endogenous and exogenous, to use biological terms, or nature and nurture, to use psychological terms. The first category comprises intrinsic properties of the rock, and the second comprises the technique used in performing the completion.

Candidates for intrinsic parameters are total organic carbon (TOC), mineralogy likely best characterized by the brittleness index (BI), number and distribution of natural fractures, organic porosity (defined in Chapter 7), hydrocarbon saturation, and matrix permeability. Although most of these could possibly be non-uniform along the horizontal, the best candidates for variability are the first three, especially the natural fracture pattern.

The problem is that the precise effect of natural fractures on production rate is not settled geoscience.[12] On the one hand, fractures of a certain size certainly augment the spread of the induced fractures. However, it is also known that some fractures act as thief zones and rob the induced hydraulic pressure by leaking away at a fast rate. The key factors involved include the specifics of size, direction, and distribution. Herein lies the problem. Current methods are not adequate to assess the in situ state. Most research in this space has been on outcrops and cores. Borehole investigative methods invariably have very low depth of penetration into the rock, and all are relatively costly for these types of wells. If the returns are there in the form of improved recovery rates, the cost may be sustainable. However, showing causality (test results guiding behavior that results in improved recovery) will be necessary to justify extra expenditure.

Microseismic providers are researching the ability to map natural fractures. Multiple surface arrays attempt to overcome the signal-to-noise ratio problem with sheer numbers. Placing the arrays in post holes approximately 100 feet below helps. However, that has additional cost. The earth close to the surface is weathered, and this layer is known to cut out signal. Being below the weathered layer would improve the signal-to-noise ratio. The principle is that the induced fractures energize the natural ones and these events can be tracked.

Part of the problem with detecting fracture patterns is that, with current methods, this occurs only after the fracturing event. Consequently, the findings cannot be applied to the same borehole. In principle, they could be used in the next horizontal, but because it is established that each one has unique heterogeneities, the utility on the next well is likely limited. However, if the microseismic findings could be correlated with borehole imaging with a high degree of confidence, borehole imaging could be used as a proxy for deeper images. This would allow selection of zones with higher productivity because the imaging log would be run in the open hole before the completion step.

Chapter 7 describes with some generality that shale is not always the same and that the prospectivity is significantly determined by the relative concentrations of three classes of minerals: clay, sandstone, and carbonates. Geoscientists have drilled deeper into this argument and nuanced it with respect to the role of natural fractures. Fig. 8.2 is a ternary diagram with characteristics of the major plays. On one vertex is clay, whose ductility makes it fracture resistant. Not shown is the belief by the authors of the figure that compositions greater than 50% clay would not be prospective. This would be a horizontal line joining the two 50's on the angled axes. Clearly, portions of the prolific Eagle Ford, Barnett, and Marcellus are out of the money.

The nuancing occurs in the other two vertices in Fig. 8.2. The left bottom vertex is represented by structures that are generally open and dominated by sandstone or feldspar. If the cracks are filled with minerals, they are of a type that does not cement to the rock walls and

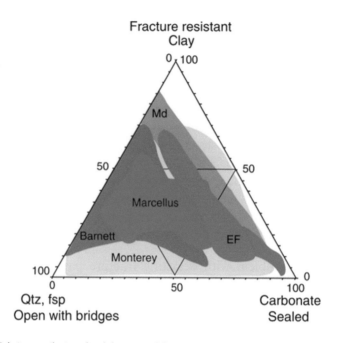

Figure 8.2 *Relative contribution of sealed, open, and fracture-resistant rock in US plays. EF, Eagle Ford; fsp, feldspar; Md, mud rock (clay-rich shale); Qtz, quartz.* Courtesy: Gale JFW, Laubach SE, Olson JE, Eichhubl P, Fall A. Natural fractures in shale: a review and new observations. AAPG Bull 2014; 98(11): 2165–2216.

consequently are prone to fracture under stress. The bottom right vertex comprises carbonates with natural fractures that are sealed to resist movement of the planes under stress.

The foregoing discussion highlights the role of natural fractures, both aiding and hindering the hydraulic fracturing operation.

METHODS TO IMPROVE RECOVERY

Using the Bit as a Sensor

The concept of using the bit as sensor dates back to the early days of MWD more than three decades ago. All methods rely on the fact that the energy required to cut rock is indicative of both rock properties and the differential pressure that the rock is encountering. This parameter is defined as the wellbore pressure less the pore pressure. Early investigators estimated pore pressure at a rough level with this technique. It enabled operators to drill safely at low differential pressures, resulting in higher rates of penetration of the bit. This is known as drilling close to balance.

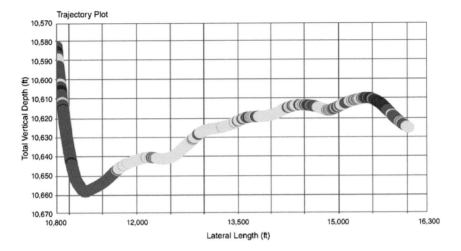

Figure 8.3 Wellbore map showing color-coded variations in hardness. Hardness is directly correlated to MSE. Courtesy: Journal of Petroleum Technology.

Today, the key measure being estimated is the mechanical specific energy (MSE). It is calculated from measurement of parameters such as weight on bit, torque on bit, revolutions per minute, and rate of penetration. This is an estimate of the energy required to excavate the rock. The measurements are made either at the surface or in a special element of the drill string near the bit. In the case of the latter, the measurements can be precise enough to yield conventional rock strength parameters such as Young's modulus and Poisson's ratio. The low-cost variant of such a service retrieves recorded data when the bit is pulled out of the hole. The raw data are transmitted to the cloud, and sophisticated data analytics allow interpreted information to be returned within hours, which is sufficient for decisions to be made affecting fracturing stages.

Measurements such as those detailed previously are still in the early stages of introduction. They attempt to correlate the measured property with the fracturing results. If a correlation is established in some measure in an area, it may be used to fracture only selected intervals. Fig. 8.3 is a plot of estimated rock hardness along the wellbore, with essentially a heat map. The cooler colors are softer and the hardness is correlated with MSE.

Analyzing Cuttings

When rock is drilled, the fragments returning to the surface in the drilling fluid are known as cuttings. There are two main types of drilling bits: roller cone and polycrystalline diamond composite (PDC).

Roller cone bits cut rock by a compressive action; PDC bits cut in shear. PDC is the most common type and delivers cuttings that are more amenable to analysis compared to roller cone.

Cuttings analysis in some form has been done for decades as a means of understanding reservoir characteristics in a minimally intrusive manner. The technique falls under the general rubric of mud logging and has been limited in specificity. The common analytical techniques are laser spectroscopy and X-ray fluorescence. The former produces an elemental distribution that indicates rock species. The latter can identify hydrocarbons. These operations are performed at the rig site and thus are quite useful in exploration to give a rough idea of rock properties.

For our purposes, a higher level of sophistication is needed. Ideally, one would like to estimate total organic carbon (TOC) and brittleness index (BI) to a reasonable resolution. The problem with cuttings analysis has largely been matching the cuttings sample to the depth from which it was derived. However, industry has become better at this, and 50-foot resolution is considered feasible. Furthermore, the estimate does not need a high degree of accuracy. For selecting the most prospective zones to fracture, a three-color map will suffice. Zones are green, orange, or red, representing levels of TOC. Similarly, one would plot BI along the wellbore and use both estimates to arrive at consensus green and orange areas. At first blush, one simply wants to avoid the red areas. Returning to the work of Miller et al.,[11] if the approximately 30% of unproductive clusters could be identified correctly, this would significantly reduce the cost of the well. If a planned 30-stage lateral had 10 stages saved, that would be a cost savings of approximately US$600,000 at depressed 2016 prices. For rough comparison of savings, the full well likely would cost approximately US$3 million.

Fifty-foot resolution is acceptable because fracturing stages tend to be 300 to 400 ft long. Within each stage, there are multiple perforation clusters. The work of Miller et al. and others[9] indicates, somewhat counterintuitively, that two clusters are better than six. However, a finer resolution of depth would be an advantage, and 30-foot resolution is likely feasible. The coarsest resolution that is effective is desirable because finer resolution means more cost for sample picking, preparation, and analysis.

In this book, we focus on two methods to exploit the means described previously. One is alluded to in Chapter 7 through the use of digital rock physics implemented on cuttings. In particular, see Fig. 7.9, which plots several parameters, including TOC and BI. Chapter 9 describes a completely different technique using DNA sequencing to find the distribution of bacteria in the rock. Through machine learning, the distribution is correlated with oil production potential of the zone. This is plotted in Fig. 9.4 and validated with tracer data.

A key question is whether the analyses will be available in time to affect the fracturing step. Currently, both methods rely on analytical operations done at central locations, not the rig site, at least at the present time. Consequently, it is reasonable to believe that the lag time is several days. In the ordinary course of events, the completion would follow immediately after the drilling, and the information would be too late. However, there is a way to obtain this information in a timely manner.

The shale industry has for some time been pursuing a practice normally done on offshore platforms: batch drilling. Multiple wells are drilled to a certain point. Then all of them are taken to the next step, and so on. Offshore, this makes sense because one can achieve economies of scale and logistics. The shale industry has increasingly employed drilling on pads, as described in Chapter 1. Because there are going to be multiple wells anyway, these can be drilled and completed in batches. There are clear economies of logistics in that identical equipment is used for a longer duration. This is particularly effective in the completion step, when materials such as steel casing, cement, and fracturing fluid are transported in bulk. Similarly, the fracturing trucks make deliveries only once, which reduces wear on roads. Whether commonly practiced or not, it is now possible to do so, and there are good arguments for economies achieved in doing so.

If batch drilling is used, all of the wells together can be drilled but not completed in 2 or 3 weeks, and the interpretations discussed previously would then be available for the completion operation. Sequential drilling and completion, by contrast, would not allow for interpretation during drilling to be implemented in the completions.

Wellbore Logging and Microseismic Monitoring

Ordinarily, the only wellbore logs will be the triple combo, as discussed in Chapter 7. The traditional parameters of gamma, resistivity, and density are unlikely to be sufficient in this setting and so usually will not be run. However, if they can be correlated with some of the more sophisticated interpretations discussed previously, they may be useful in combination or by themselves. No doubt this will be case specific.

Borehole viewers are much more useful because they can identify natural fractures, at least to the extent that the fractures intersect the borehole. Because most natural fractures are believed to be vertical, and the wellbore is horizontal, the chances for intersection are good. An interesting combination would be the use of viewers with microseismic measurement. The latter measures only during the fracturing event and so is temporally displaced from when the borehole device is run. Nevertheless, if formation fractures can be correlated with the imaging tool, then the tool would be useful in its own right. The advantage is that the results would be obtained in time to guide zonal selection.

Microseismic monitoring lights up all areas of deformation in the rock. Because induced fractures are believed to activate natural fractures, they can be identified by microseismic monitoring. Since this entire fracture network can be presumed to be the active conduit for hydrocarbons, if properly modeled and characterized, it may be used for flow simulation. The most useful flow simulator would be one with closed-form computation, which can produce a simulation in minutes. There is still the problem that all of this is post facto relative to the fracturing event. However, if indeed the results are available in minutes, it is not inconceivable to inform a successive zonal operation. High-speed computation would also enable some "what if" experiments. This is in the realm of future research.

Secondary Recovery

This is the term used for recovery of additional hydrocarbon after the first attempt. In conventional reservoirs, it is generally accomplished by drilling a well proximal to the old depleted one. Then water with or without chemicals such as surfactants is introduced into this injector well, and it sweeps across the reservoir, pushing the remaining oil ahead of it into the original producing well. In cases in which the oil is bound to the rock by divalent ions, care is taken to ensure there is a high proportion of monovalent ions in the water flood. With ion exchange, the divalent ion is released and takes the bound oil with it. Sometimes natural gas is alternated with the water (water alternating gas (WAG)), this being more of a practice when the oil is heavy.

The WAG technique is unlikely to work in unconventional reservoirs because the low permeability does not permit a flood in the classic sense. The most likely approach is to inject CO_2 or some other miscible fluid into the producing well and shut it in for a period, after which the well is put on production. Industry refers to this approach colloquially as huff and puff, and one of the keys to the modeling is the appropriate soak time. The technique is commercially used for heavy oil where the stimulating ingredient is steam. Whatever the mechanics of getting the miscible fluid in, the leading candidates are natural gas and CO_2. The latter is not easily available. Natural gas at US\$2 per MMBTU (early 2016 price) is a good exchange for oil at US\$42 per barrel ($\sim$US\$7 per MMBTU). However, there are only sporadic reports of attempts to employ this substitution. There is an ongoing study at the Colorado School of Mines.

Refracturing

In the opinion of many, including one of the authors (VR), refracturing is the most promising avenue to enhancing recovery.[13] This does not qualify as secondary recovery because the methods used are virtually the same as those used the first time. It involves going back into a well after production has gone near asymptotic and repeating the fracturing step at discrete intervals. One problem with this method is that it has had spectacular successes and failures. Also, most use has been in gas wells. To date, the principal driver for refracs is that the original completion was inadequate in some way. Without going into details, more than a decade ago the technology was such as to either damage the formation or simply not drain the reservoir effectively, usually because of improper use of proppants. One would go back into the same zone and do it correctly. The other case is where a good well simply has depleted. One then fractures fresh rock proximal to the zone. To achieve this, the original perforations must be sealed during the operation. The most recent technology for doing that is the use of degradable polymers such as polylactic acid or cellulose acetate, which degrade over time and at rates dependent on temperature. The formulations, especially of the acetate, can be designed to degrade at different rates under different ambient conditions. The intent is that after the refracs are complete, the original perforations are opened up again for any little production they may provide.

A consortium in the Eagle Ford performed a study of six oil-producing reservoirs.[14] The primary goal was to regain 60% of the original production while spending no more than 40% of the original cost. Fig. 8.4 shows that this goal was largely met. The production increases ranged from 31% to 76%. The associated cost was 70% less than that of drilling and completing a new well. The method used a new degradable fiber for diversion. As a detail, the consortium also used degradable fiber-laden fracturing fluid to increase the proppant carrying capacity, ensuring deeper penetration. Eventually, the fiber degrades, leaving the proppant in place in the fractures.

Current thinking revolves around enhancing already good producers. Keep in mind that in each 400-foot interval, there are only two or three clusters, primarily for reasons of hydraulic efficiency to get the most penetration. This still leaves considerable rock untreated because lateral drainage is usually poor unless there are significant natural fractures in the correct configuration. Consequently, the best

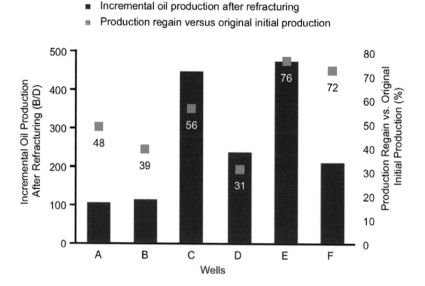

Figure 8.4 Production gains through refracturing in the Eagle Ford. Figure courtesy JPT; original data courtesy Schlumberger.

strategy for refracs is to identify the best producing zones, not the poor performers, as had been the practice. In terms of analytical methods, this brings us squarely back to the same technologies as those used to identify lateral heterogeneities with respect to production rates. In other words, just about every method identified previously in this chapter applies toward selecting refrac candidate spots.

CONCLUSION

In summary, the most promising route to improving profitability of shale oil and gas resources is increasing net recovery. Certainly, continued emphasis on reducing costs of operations is valid, but much has already been achieved here. In cost per barrel, attacking the denominator is more long-lived. This is in part because the cost gains have come in two forms. One is operational efficiency such as pad drilling, which will persist. The other is plummeting service costs. For example, fracturing costs have declined by more than half since early 2014. These costs will inevitably creep back up when there is a return to robust volume. Consequently, increasing recovery is the surest long-term road to lower break-even cost and resilience against future oil shocks.

REFERENCES

1. Gale JFW, Holder J. Natural fractures in some US shales and their importance for gas production. *Petroleum Geology Conference Proceedings* 2010;7:1131–40.

2. Jacobi DJ, Gladkikh M, LeCompte B, Hursan G, Mendez F, Longo J, et al. Integrated petrophysical evaluation of shale gas reservoirs. *SPE 114925-MS*. The Woodlands, TX: Society of Petroleum Engineers. <www.onepetro.org>; 2008. [accessed 15.07.16].

3. Gale JFW, Holder J. Natural fractures in the Barnett Shale: constraints on spatial organization and tensile strength with implications for hydraulic fracture treatment in shale-gas reservoirs. The 42nd U.S. Rock Mechanics Symposium (USRMS), 29 June–2 July, San Francisco, CA, 2008.

4. Barton CA, Castillo DA, Moos D, Peska P, Zoback MD. Characterizing the full stress tensor based on observations of drilling-induced wellbore failures in vertical and inclined boreholes leading to improved wellbore stability and permeability prediction. *APPEA J* 1998;29–53.

5. Warpinski NR, Teufel LW. Influence of geologic discontinuities on hydraulic fracture propagation. *SPE-13224-PA*. The Woodlands, TX: Society of Petroleum Engineers. <www.onepetro.org>; 1987. [accessed 15.07.16].

6. Warpinski NR. *Hydraulic fracturing in tight, fissured media. SPE-20154-PA*. The Woodlands, TX: Society of Petroleum Engineers. <www.onepetro.org>; 1991. [accessed 15.07.16].

7. Mason S. The role of natural fractures in the process of hydraulic fracturing. Research Paper, University of Louisiana at Lafayette; 2014.

8. Schein G. Hydraulic fracturing in gas shale plays: are they all the same? <http://barnettshalenews.com/documents/GarySchein-FtWorthBusinessPress-ShaleCompletions-022908.pdf>; 2008 [accessed 15.07.16].

9. King GE. 60 years of multi-fractured vertical, deviated and horizontal wells: what have we learned? *SPE 170952-MS*. The Woodlands, TX: Society of Petroleum Engineers. <www.one-petro.org>; 2014. [accessed 15.07.16].

10. Rao V. *1984 and beyond: the advent of horizontal wells. SPE-1007-0118-JPT*. The Woodlands, TX: Society of Petroleum Engineers. <www.onepetro.org>; 2007. [accessed 15.07.16].

11. Miller CK, Waters GA, Rylander EI. *Evaluation of production log data from horizontal wells drilled in organic shales. SPE 144326-MS*. The Woodlands, TX: Society of Petroleum Engineers. <www.onepetro.org>; 2011. [accessed 15.07.16].

12. Gale JFW, Laubach SE, Olson JE, Eichhubl P, Fall A. Natural fractures in shale: a review and new observations. *AAPG Bull* 2014;**98**(11):2165−216.

13. Jacobs T. Renewing mature wells through refracturing. *J Pet Technol* 2014; April.

14. Acock A, Clark B. Sequenced refracturing technology improves economics in unconventional plays. *J Pet Technol* 2015; September.

Subsurface DNA Sequencing: A New Tool for Reservoir Characterization

Evaluation of unconventional reservoirs using the seismic, logging, and associated petrophysical techniques developed over the years for conventional reservoirs faces many challenges, as described in Chapter 7. That chapter also discusses the digital rock physics approach to addressing these inadequacies. In particular, rock layers are normally considered homogeneous, and heterogeneous rock layers are proving intractable. A horizontal well in a given stratum often evinces strikingly different petrophysical characteristics in a matter of meters. These different characteristics lead to variable production performance in the conventional practice of geometrically designed fracturing, where the fracturing zones are usually of uniform length and spaced at predetermined intervals. Up to 30% of the perforation clusters designed using these strategies are nonproductive.[1]

There are two main approaches to acquiring the information needed to selectively fracture more prospective zones. One is to estimate rock properties while drilling, essentially using the bit as a sensor (see Chapter 8). The other is to evaluate drill cuttings (see Chapter 7) or produced fluids. Both approaches are nonintrusive to the drilling and completion environment, and using them does not cost operational time. In the low-cost environment of shale oil and gas drilling, these considerations are important for widespread usage. The interpreted information must be available in time for operational decisions, and the increasingly popular practice of batch drilling facilitates timeliness. In this concept, multiple wells are drilled but not completed. The fracturing step is conducted in all the drilled wells together in a batch. This lengthens the time span between retrieving and analyzing cuttings and the fracturing step that the analysis informs, compared to the conventional method of completing immediately after drilling. The digital rock physics approach described in Chapter 7 requires approximately 10 days to return the interpreted result.

Sustainable Shale Oil and Gas. DOI: http://dx.doi.org/10.1016/B978-0-12-810389-0.00009-7

MICROBES IN ROCKS AND ELSEWHERE

For more than 60 years, it has been known that life is possible in severe environments. In the 1960s, microbes were discovered living in the hot springs of Yellowstone National Park,[2] even in boiling water. The hot springs also yielded a species of bacteria, *Thermus aquaticus,* from which an enzyme was later found that could copy DNA at high temperature.[3] This led to a revolution in laboratory techniques called the polymerase chain reaction (PCR) that allowed exponential amplification of DNA by successive cycles of copying and heating to separate the strands, and it enabled much contemporary genetic research. Researchers also found life in the deep ocean at deep-sea sulfide vents,[4] which have temperatures upwards of 113°C (235°F) and pressures of 100−300 bar (1450−4351 pound-force per square inch). Because these temperatures and pressures interact with the colder ocean depths, some liquefied minerals become solids and the properties of water can change. Microbes have adapted to living without light and oxygen in this environment, surviving on reduced chemical compounds such as sulfur, methane, iron, and manganese. In this environment, bacteria have even been found that can practice photosynthesis (whereby energy for growth and metabolism is made from light) not from sunlight but from geothermal radiation produced from the deep-sea vents that includes wavelengths that can be absorbed by the photosynthetic pigments.[5]

Deserts also yield important insights into the extreme environments in which microbes can live. Surfaces of desert rocks (so-called desert varnish) harbor microbial communities that use trace amounts of metals from the atmosphere to survive.[6] Some microbial organisms live in arid conditions and dry out until water reappears, remaining dormant for years. Microbes are even found meters under the desert surface in high-salt environments.[7]

Microbes and Oil Spills

The Exxon Valdez oil spill off Prince William Sound in Alaska in 1989 saw the first full-scale use of biological remediation for an oil spill. Biological remediation consists of enhancing the growth or activity of microbes already present in the environment, or adding microbes to the environment, to increase microbial degradation of contaminating compounds. Oil from the Exxon Valdez floated on the water and quickly came ashore. Chemical dispersants were used in an attempt to break it up, but they largely failed. To clean up the shoreline oil, nitrogen fertilizers were applied to enhance native microbial degradation.[8] The result was the largest bioremediation effort ever taken, with more than 107,000 pounds (48,600 kg) of fertilizer applied during a 3-year period.

In 2010, the Deepwater Horizon oil platform exploded and sank, releasing high-temperature and high-pressure light crude oil and gas from the wellhead at approximately 4265 feet (1300 meters). The platform was nearly 40 miles off the coast of Louisiana. Due to the major threat to Gulf of Mexico community economies, officials approved the use of the dispersant Corexit applied at the deep wellhead with the goal of keeping oil beneath the surface and allowing greater microbial remediation to prevent it from going ashore. Dispersion was also aided by the high temperature and pressure of the oil released from the wellhead, which reduced the physical size of the oil particles and aided microbial remediation. During the spill, researchers studied the progression of microbes found in the oil plume using culture-independent high-throughput genetic sequencing.[9] These studies, which covered a much broader segment of the microbial community than can currently be grown in the laboratory in culture, showed the progression of microbes that dominated remediation in its various stages,[10] with early stage increases in microbes already present in the water column that could degrade n-alkanes and cycloalkanes[11] and later stage increases in microbes that could degrade cyclic aromatics (which contain benzene rings).[12]

Dispersants were intended to break the oil into smaller droplets. Critics believed that the smaller droplet size would lead fish and larger life-forms to consume more of them and become more highly contaminated. British Petroleum has been silent on their reasons for using dispersants, but it is likely that the company believed that bacterial action would be enhanced by the greater surface area-to-volume ratio of the smaller drops. Nobody disputes the following three premises about the bacteria: They do consume oil, they multiply in the presence of food, and they die off when the food supply goes away. Critics believed that the dispersant-laden oil would sink below the surface and would then be unavailable for bacterial action because of the colder waters with less oxygen at greater depths. The following was the key question: Would microbial species that could biodegrade the oil be found in the deeper environment?

Hazen and colleagues at Lawrence Berkeley Laboratories found psychrophilic (cold-temperature) Gammaproteobacteria closely related to oil-consuming species near the blowout.[9,13] Scientists were concerned that even if microbes capable of biodegradation existed in this environment, they would deplete the available oxygen. This oxygen depletion would then lead to dead zones, water in which life cannot be supported. However, the study found that the newly discovered creatures did their job with minimum oxygen consumption. Oxygen saturation was 59% within the plume compared to 67% outside it, which was an acceptable decrease.

Although reservoir rock is an extreme environment, the existence of bacteria there is not surprising, because they live in many other extreme environments. Bacterial species have important roles in hydrocarbon maturation. Oil and gas are created when dead plants and animals decompose under high pressure and temperature, by both biotic (of biological origins) and abiotic (purely chemical) processes. Geochemical techniques have shown that some chemical compounds in oil from igneous and sedimentary rocks show a biotic origin. Microbes degrade heavy (longer carbon chain-containing) hydrocarbons to create lighter hydrocarbons, showing that microbes are important in producing the more desirable lighter oils.[14] This microbial capability is featured in a later discussion about microbially enhanced recovery. Microbes have also been shown to degrade polycyclic aromatic hydrocarbons under anaerobic conditions,[15] providing further evidence that biodegradation of organic materials by microbes forms the biotic portion of oil in the subsurface.

Heavy oil, which is characterized by a predominance of large molecules, is believed by some to have been formed by the action of certain species of bacteria. These species consume hydrocarbons as food and are known to prefer smaller molecules. The product left behind after this process is oil rich in large molecules, with higher specific gravity and greater viscosity, and therefore characterized as heavy oil.

MICROBES AS FORENSIC MARKERS

This chapter introduces the concept of using microbial distributions as indicators of hydrocarbon potential. Because this is essentially a forensic exercise, we summarize how microbes are used in forensics in other settings.

The human body has more microbes on and in it than cells containing the human genome. These human-associated microbes are highly specific to individuals, forming a microbial fingerprint that identifies one individual from another along with the objects touched by that individual. Researchers from the University of Colorado demonstrated that humans transfer their microbial fingerprints onto their keyboards as they type. These fingerprints accurately identify which human typed on which keyboard.[16] Similarly, a study of microbiome samples taken from phones and shoes showed that not only do individuals have unique microbial fingerprints but also even the location where the sample is taken from each person has a unique microbiome. Shoes from each person were discerned by matching their microbial fingerprints.[17] Other work at the University of Colorado showed that people who lived together shared microbes with one another (i.e., each family and each household, including households containing unrelated individuals such as lodgers, have microbial signatures), but individuals could still be identified from their unique microbial fingerprints.[18] Enhanced resolution may be possible by using a technique called shotgun metagenomics. This technique provides a complete inventory of genes and higher taxonomic resolution than the marker gene analysis in the studies described previously, which are nonetheless highly accurate.[19] It is this ability of microbes to be transferred between media and the high detail of the resulting data that enable microbial DNA sequencing to provide useful, actionable insight in many oil and gas applications.

The early applications of forensics in oil and gas were in the use of microbes as markers in exploration. Underwater seeps of oil and gas were located by identifying species of bacteria that oxidized the hydrocarbon molecules. The more mobile molecules methane, ethane, propane, and butane are likely to make it to the surface from the reservoir. Bacteria that can oxidize these molecules reside near the surface and are detected. Methane can also be from biogenic sources, so the relevant bacteria (methylotrophs) are imperfect indicators of hydrocarbon seeps because they may instead be feeding on biogenic methane from the surface. Species that oxidize larger hydrocarbon molecules that are unlikely to be produced biogenically in the subsurface are more reliable indicators.

A more sophisticated use of bacteria in diagnostics is found in studying thermophiles and other extremophiles in oil and gas settings.[20] Thermophiles survive in temperatures exceeding 50°C. Most hydrocarbon breakdown occurs at temperatures greater than 60°C. Extremophiles thrive at extremes of temperatures, pressure, salinity, pH, or other environmental conditions. Extremophiles are associated with hydrocarbon accumulations, possibly because they feed on the hydrocarbon molecules. Consequently, they are used as markers in oil and gas exploration. This diagnostic technique can be applied to any subsurface material, including cores, produced oil or water, and shallow sediment near seeps.

Microbially Enhanced Oil Recovery

This chapter focuses on the analysis of microbial communities resident in petroleum reservoirs and so can be classified as forensic. A completely separate endeavor is to stimulate growth of desirable microbes or add to the microbial communities in order to manipulate the molecules in the reservoir. This is classified as microbially enhanced oil recovery (MEOR). That is a misnomer because natural gas can also be a target and because in some cases, unlike most EOR methods, the net recovery is not increased but instead a fluid with more desirable attributes is recovered. Because of the distinctively different mission of MEOR over forensics, this discussion is relegated to a text box.

Until shale oil came on the scene, heavy oil was believed to be the next major source, and it still represents an enormous resource base. The "heavy" designation is due to the high specific gravity of the fluid because of the predominance of very long-chain hydrocarbon molecules. The operational impact of heavy oil is that the high viscosity makes movement through the reservoir and lifting out of the reservoir difficult. Transport in pipelines requires thinning, usually with lighter molecules and sometimes with heat. The costs of recovery and transport are therefore higher than those for light oil.

Heavy oil also poses environmental challenges. The weight of the molecules and the energy expended to overcome the viscosity result in a higher carbon footprint than for light oil. This has been the singular stated reason for opposition to the Keystone XL pipeline from Canada to the United States.

All of this could be overcome if the molecules were manipulated in situ to reduce the carbon footprint. Simply stated, the objective is to decrease the carbon:hydrogen ratio by cleaving the largest molecules. Today, this chemical processing occurs in refineries, but MEOR attempts to achieve it microbially in the subsurface,[21] although catalytic methods are also possible.

The search for organisms that convert heavy oil to a lighter version is ongoing. Statoil has reported gathering a number of organisms with an ability to break down heavy oil molecules.[22] Under experimental conditions, complete conversion of heavy oil to lower viscosities has been achieved within 2 or 3 days of adding specific bacteria.

The most degradable compounds in crude oil are straight chain aliphatics (alkanes), followed by branched isomers and cyclic hydrocarbons. Several investigators have concluded that anaerobic degradation of crude oil in subsurface reservoirs under methanogenic conditions explains the hydrocarbon compositions seen in degraded oils.[23] Fig. 9.1 plots the results when undegraded crude oil is subjected to methanogenic conditions. The decrease in the population of alkanes is striking

(nC_7 to nC_{34}), commensurate with an increase in methane production. The authors suggest that the initial bacterial breakdown of the large molecules results in acetic acid.[23] Two possible pathways then take the acetic acid to methane.

These were laboratory studies. Note also the significant time involved, although for in situ molecular manipulation, long soak times may not be deterrents. Unlike terrestrial chemical processing, in which capital utilization drives the need for shorter processing times, industry is likely to take the trade-off of time versus oil quality in situ. Examples of such reasoning already exist. Royal Dutch Shell researched,[24] and may implement, a long-term soak approach to recover oil from oil shale deposits (oil shale is essentially kerogen, not to be confused with shale oil, which is light oil).

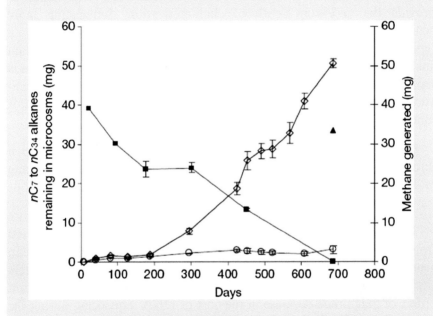

Figure 9.1 Microorganism-mediated depletion of n-alkanes and production of methane. Courtesy: *Nature.*

WORKFLOWS FOR DNA SEQUENCING IN THE SUBSURFACE

Before DNA sequencing, microbiologists could only study microbes that could be cultured in a lab (the term used is *culturable*). The conditions for growth therefore had to be determined before any further work could be completed. Microbiologists could see how the colonies grew, how they interacted with each other, and how particular bacteria favored certain energy sources based on their ability to grow. Unculturable bacteria, on the other hand, cannot be physically manipulated in the lab in the same way. As such, researchers often restricted their studies to a small fraction of the microbial community that could be more easily manipulated.

It took the discovery of endonucleases (essentially genetic cutting shears) and genetic amplification techniques in the 1980s for scientists to begin to characterize the unculturable.[25] However, unraveling more of the lower-abundance microbes present (the "rare biosphere") and fully characterizing a community would take nearly another two decades for modern barcoding and next-generation sequencing to provide the necessary depth of characterization. Sequencing more genetic material from samples yielded a data analysis problem, as a typical project scaled from dozens to millions of sequences. The data analysis of millions of sequences to yield useful outcomes requires a separation of signal from noise, replicated study designs, and techniques for identifying statistical links between microbial taxa and/or functions and parameters of interest.

DNA sequencing of the subsurface has therefore been built upon three fundamental technological advances: next-generation sequencing of DNA, scalable computing and storage, and software innovations for handling DNA. All are described in this chapter.

The human genome was first sequenced during the 10-year period from 1990 to 2000 at a cost of approximately US$3 billion. A human genome today can be sequenced over a few days for approximately US$1000, enabled primarily by next-generation sequencing. That several orders of magnitude decrease in the cost of DNA sequencing during the past 15 years allows the technique to drive innovation across many disciplines, including health care, agriculture, and now energy. In the same time frame, and with the advent of cloud-based systems, the cost of computing and data storage has also decreased significantly, and access has become ubiquitous. The application of DNA sequencing to the oil and gas industry and, for that matter, personalized medicine would not be possible without the removal of these economic barriers.

Fig. 9.2 shows the workflow, beginning with sample acquisition and ending with actionable interpretation. Sample acquisition is the first step and involves service personnel (these individuals are known as mud

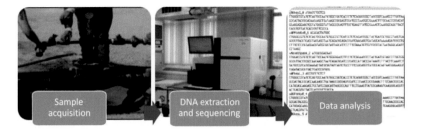

Figure 9.2 Workflow in DNA sequencing of the subsurface. Courtesy: Biota Technology, Inc.

loggers) collecting well cuttings directly from the shakers, which are simple filters that separate solids from liquids returning to the surface. Similarly, oil samples can be collected directly from the wellhead. A compelling aspect of this method is that the sample acquisition is essentially nonintrusive. No special downhole tools are required, unlike the logging methods described in Chapter 7.

The ease of collection of these oil field sample types reveals a major strength of DNA sequencing as an investigative technique. Operators can evaluate DNA sequences over time, with no downhole tools, using their existing operational workflows. With a noninvasive sample acquisition procedure that disrupts operations only minimally, DNA sequencing enables high-density time series to be collected and analyzed. DNA sequencing has numerous other advantages over earlier techniques. For example, chemical tracers provide data for only a few months and then dissipate, but microbes and their DNA can be found for the life of the well. DNA sequences can thus be used to profile the microbes found in the formations of a well once and then monitor active fracture height and production profiling over the life of the well. Samples can be collected without disruption on timescales ranging from several times a day to once every 6 months, depending on operator needs and budgets.

SAMPLE ACQUISITION

Well cuttings, rock, drilling mud, fracturing fluids, produced water, flowback, and oil can all be sampled, depending on the desired application. Given the sensitivity of genetic sequences produced, engineers must be very careful in collecting samples in order to minimize contamination in the entire workflow from collection through lab work, such that biological noise is kept to a minimum. To detect and, where possible, correct for inevitable contamination, samples from the site are also collected to track local contamination sources. Samplers wear gloves, and samples are collected in standard sterile containers that are free of DNases and RNases (enzymes found naturally on skin and in bacteria that degrade DNA and RNA, respectively). Samples are shipped cold in coolers back to the lab. Per-sample information, often called sample metadata, is recorded about each sample. This metadata includes well name, depth of the sample, sampler, type of sample, collection date and time, and other relevant information for the analysis (including analysis of possible confounding variables).

At the lab, samples are stored at $-80°$ or $-20°C$ until ready for DNA extraction. Some properties of the samples, including high salinity and increased metals, can necessitate special techniques for DNA extraction. Because a key step in the microbiome pipeline is amplification of DNA corresponding to genes of interest, in which individual DNA molecules undergo PCR amplification to create millions of copies of the original DNA strands,[3] any contaminants left with the DNA that inhibit the DNA polymerases will hamper amplification. Because of the large amounts of DNA ultimately produced by this process, the goal is to find extraction methods that are chemically and physically similar across sample types rather than to optimize yield from each sample while introducing sample-specific biases that are difficult to control for. Similarly, enzymes and primers used in amplification protocols must be kept consistent to compare results across different sample sets. Many investigators have observed changes in the microbes inferred to be present in each sample with changes in extraction methods, amplification primers, and/or data analysis methods. To confidently compare results across sample types and studies, consistent protocols must be employed. This includes next-generation sequencing protocols, which have to be modified for microbiome work and the potential low diversity that samples may contain. A good place to start for oil and gas samples is the protocols developed and used by the Earth Microbiome Project.[26]

DATA ANALYSIS

Recent discoveries about the microbiome would not have been possible unless the decrease in sequencing costs had progressed in tandem with the application of a "big data" mindset. For example, one of the first tasks in microbiome research is to compare DNA sequences to reference databases of previously identified and described bacteria. In the early days of sequencing, when a study might yield dozens of sequences, this task could only be carried out serially (one sequence after another), and each sequence could be directly compared to the necessarily small reference database. However, the explosion of sequences quickly meant that this old approach was no longer computationally feasible. Operational taxonomic units (OTUs), or genetic sequences grouped by a pairwise percentage of identity in the DNA sequence (e.g., 97% OTU is approximately species level), can be constructed in different ways. One recent solution is the so-called open reference OTU picking protocol, which parallelizes the direct comparison of sequences to the database (closed reference) while also allowing for de novo OTUs to be identified. This protocol allows researchers to compare billions of sequences against reference databases, preserving stability of matches to known sequences across studies while also allowing newly discovered sequences to be incorporated.[27]

The key data analytical engine used is the open source QIIME,[28] described as follows by High-Performance Computing at the US National Institutes of Health:

> QIIME (pronounced "chime") stands for Quantitative Insights into Microbial Ecology. QIIME is an open source software package for comparison and analysis of microbial communities, primarily based on high-throughput amplicon sequencing data (short genetic sequences that are amplified such as SSU rRNA) generated on a variety of platforms, but also supporting analysis of other types of data (such as shotgun metagenomic data). QIIME takes users from their raw sequencing output through initial analyses such as OTU picking, taxonomic assignment, and construction of phylogenetic trees from representative sequences of OTUs, and through downstream statistical analysis, visualization, and production of publication-quality graphics. QIIME has been applied to single studies based on billions of sequences from thousands of samples.

The QIIME method includes quality control for sequences.

MICROBIOME DATA ANALYSIS MEASURES

Alpha Diversity

This concept relates to the number of different types or taxa of microbes within each sample—its species richness. Samples can also be characterized by the number of observed OTUs, or the number that would be predicted to be present if sampling had been exhaustive. The most basic alpha diversity metric is to simply count how many taxa or OTUs are observed in each sample. Samples with higher alpha diversity will have more taxa. Researchers often compute alpha diversity curves, where the alpha diversity metric is calculated for each given rarefaction depth (a smaller number of sequences randomly chosen as a subset of the total sequences). For instance, a sample with two sequences could contain either one or two different OTUs; a sample with three sequences could contain either one, two, or three OTUs; and so on. At some level of rarefaction, however, an asymptote will form, where many more sequences need to be examined to uncover a single new member of the community. This rarefaction level indicates that most of the microbial diversity of the sample has already been sampled. Other alpha diversity metrics, such as Chao1, are abundance-based coverage estimators that estimate the total number of species (or OTUs) from the observed counts. Whichever metric is chosen, alpha diversity can be compared across samples and related to other metadata to test for correlations. For example, comparisons of alpha diversity between sample groups taken at different temperatures may show that fewer kinds of microbes can survive at higher temperature, leading to a lower-diversity community. Note that for some applications, deep sequencing might be required, in which hundreds of millions of sequences are produced for a single sample rather than millions of sequences split across hundreds of samples.

Beta Diversity

This is a measure of the diversity between samples rather than within a single sample (which is alpha diversity). The most basic beta diversity metric might simply count the shared OTUs between two samples. This procedure is carried out between all pairs of samples in the data set of interest, and the results are stored in a matrix. A weighted version includes the abundance of each OTU. One extremely robust and useful beta diversity metric that can often tease out biologically meaningful differences between samples is called UniFrac.[29] UniFrac includes phylogenetic information when comparing two samples by calculating the fraction of the total phylogenetic tree that is unique to one of a pair of samples, as opposed to being shared between the two samples. Samples that contain phylogenetically similar OTUs (i.e., samples with communities that share common branches of the tree), but not the same OTU, are thus said to be more similar to one another than samples that contain OTUs that are more separated along the tree. UniFrac values range from 0 to 1, whereby values closer to 0 indicate two communities that are more similar and values closer to 1 indicate two communities that are more dissimilar. UniFrac distances also can be used to do classical statistical analysis when the distance matrices are compared by such nonparametric methods as the Mantel test.[30]

Group Means

Classical statistical analysis is used to determine whether the mean counts of a specific taxa are higher in one group than another. For example, the researcher might wish to know whether water flooding has introduced any new microbes to the oil and water communities. This would be done by comparing the means of OTUs of interest before and after well maintenance. Because the underlying distributions are not normal, a nonparametric test such as the Mann–Whitney U is best to determine if two samples originate from the same population distribution. However, even these nonparametric tests do not take into account some of the statistical features unique to microbiome data, such as sparsity and compositionality, and customized techniques such as analysis of composition of microbiomes that do take these features into account are strongly preferred because the false discovery rates of standard techniques can be surprisingly high even when the nominal false discovery rate is set at a standard value such as <0.05.[31]

Source Tracker

Another powerful application is source tracking. A researcher might profile the microbes living in the oil in three distinct formations within a field and want to assess the percentage allocation of each of those three sources in a commingled well that produces all three types of oil. This type of mixture modeling can easily be done with the feature-rich data of the microbiome by using probabilistic methods.[32] Furthermore, Source Tracker uses Markov Chain Monte Carlo on Dirichlet priors to estimate an unknown community, or what microbes were present in the commingled oil that likely came from some source other than one of the distinct formations. In this way, microbial source tracking has several advantages compared to other methods of production monitoring, such as comparing gas chromatography–mass spectrometry spectra, where every end member that could possibly be included in the commingled well must be profiled.

APPLICATION OF DNA SEQUENCING IN THE SUBSURFACE

Stage-Level Application: Sweet Spot Identification and Production Profiling

Chapter 8 described one shortcoming of current modes of operation—geometrically spaced fracturing zones. Heterogeneities along a lateral result in variable production along its length, especially when it was assumed to be homogeneous. A 30-stage completion could have as many as 10 zones that are relatively unproductive. The ability to identify the more productive zones is called sweet spot identification. Sweet spot identification would result in substantial cost savings by avoiding fracturing unproductive zones.

Per stage production levels are related to oil contained in place at the stage, the mechanical properties of the rock relative to fracturing potential, and the success in execution of the fracturing step. Methods for assessment of mechanical properties were described in Chapter 8. Analyses of well cuttings by geochemical techniques have not been definitive enough to persuade operators to move away from simple geometric design of fracturing zones.

A case study shows that DNA sequences can identify stages with a higher likelihood of production. The presence or absence of oil-associated microbial DNA signatures in each well cutting can be assessed through comparison with a database of oil-associated microbes and their DNA sequences. Each depth, and therefore stage, can be given an oil potential score, where higher values indicate higher potential for oil production. Fig. 9.3 shows oil potential scores applied to one horizontal shale oil well in the United States. Chemical tracers were independently used to determine producing stages during the first 3 weeks of production. As the figure indicates, the DNA sequence-based fingerprint method has a 90% correlation with the chemical tracer results. These data were generated independently (blinded to each other), validating the DNA-based method against an established technique. The general finding from both methods was that the heel was less productive than the toe. Interestingly, the DNA-based method was able to match tracers in recognizing that stage 6 was below the mean, despite being placed in the more productive toe section.

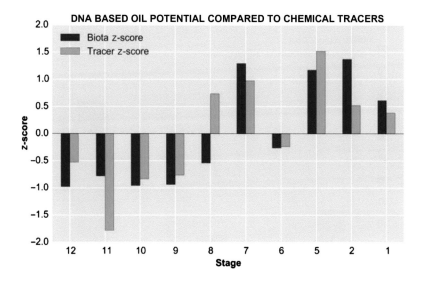

Figure 9.3 A 12-stage horizontal (stage 12 at heel, stage 1 at toe) showing z-scores of production values for tracers compared to a z-score of oil contribution scores for DNA. Per stage values were converted to z-scores to allow direct comparison between the two methods' scales. Courtesy: *Journal of Petroleum Technology*; original source Biota Technologies.

Well-Level Application: Production Profiling Across Several Formations

Hydrocarbons are sometimes produced simultaneously from more than one formation and the fluids are comingled. This is especially common in offshore conventional wells. For these wells, the operator wants to know the relative contribution from each interval. Of particular interest is whether relative contributions change over time. In offshore multilateral wells, production from individual zones can sometimes be modulated. In their most sophisticated embodiments, these are known as smart or intelligent wells. This ability would be used if the relative contributions were seen to change.

Repeatable DNA signatures are found in the produced oil and well cuttings within a single formation and across several formations. In one pilot, oils from three different formations were profiled by sequencing the DNA from in-field, single-interval wells. Using these end-members, production profiling was run on a comingled well producing from all three intervals. The results from DNA-based production profiling were independently compared to production profiling assessed through geochemistry by a major oil field service company (Fig. 9.4). The analytical mixture modeling using DNA sequences

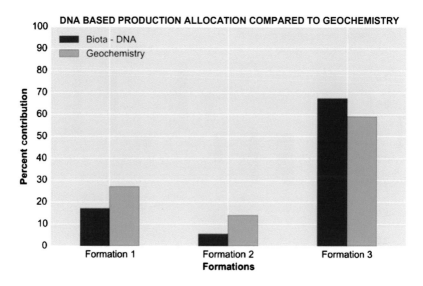

Figure 9.4 Geochemistry production profiling in a pilot well compared against DNA sequence mixture modeling. DNA sequences correctly determine the ranking of dominantly producing formations. Courtesy: Biota Technology.

correctly determined the rankings of the three intervals and matched geochemistry within ± 10%. Unlike the geochemistry-based method, end-members need not be supplied every time a DNA-based production profile is requested. The formations can be profiled with DNA sequences once, and this database of formation information is used for all subsequent analyses. This is especially relevant because single-interval zones may not be available over time when operators decide to comingle their wells with new completion designs.

Fracture Height Estimation

Shale reservoir development often entails multiple horizontal wells spaced from each other in the vertical direction. These are known as stacked laterals. Understanding the effective fracture height within a stratigraphic unit is important. It informs the vertical drainage area of the lateral, which helps define reservoir communication to improve well placement and spacing decisions in stacked lateral plays.

Pilot studies have shown overlap in DNA sequences between well cuttings and produced oil from the same wells. The shared microbes between cuttings and oil can indicate which formations, profiled with well cuttings, are producing the oil. This is a relative contribution, identifying depths with higher or lower production. The depths above

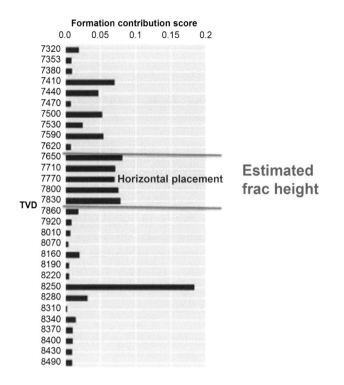

Figure 9.5 Determining the most likely depths of production, and therefore fracture height (orange lines), by modeling well cutting DNA sequences against a well's oil DNA sequences, compared to the actual placement of the horizontal well (indicated in red). Courtesy: Biota Technology.

and below where a horizontal was placed can serve as a proxy for determining fracture height because DNA from this range of depths was also found in the produced oil. To validate this technique, well cuttings were analyzed from a vertical pilot well and oil from the producing lateral. With the analyst blinded to the actual vertical placement of the horizontal, the most likely depths of oil production were determined. When the horizontal well's landing depth was revealed, it fell squarely in the middle of the estimated fracture height window (Fig. 9.5). For this pilot, well cuttings were acquired every 30 feet. More frequent acquisitions could improve the precision.

Possible Application in Conventional Plays

Sophisticated logs can be run on conventional wells, especially in deep water. The role for DNA-based investigations will be more nuanced. Because these prospects are not the subject of this book, we do not explore this matter except to point out their potential in exploration and basin definition. There is evidence that DNA signatures remain intact over time in cores. If this is confirmed, core libraries could be revisited to add this layer of interpretation. The high-resolution nature of DNA, more than 1000 times that of geochemical techniques, might be useful to improve our understanding of reservoir and basin connectivity.

Given the noninvasive nature of the diagnostic, 4 D measurements can be taken across a well's life cycle. Thus, DNA measurements may serve as a tool for continuous well health. Akin to a periodic health checkup by a physician, such life cycle monitoring of a well's DNA may enable proactive intervention before major production issues emerge. Because the microbes that reside in the oil and water components of produced fluid are materially unique, DNA can serve as a highly specific tracer to monitor secondary recovery operations such as water flooding programs and provide a unique window into a reservoir's oil and water contact point across time.

In summary, DNA sequencing shows considerable promise as a tool for characterizing oil and gas reservoirs.

REFERENCES

1. Miller CK, Waters GA, Rylander EI. Evaluation of production log data from horizontal wells drilled in organic shales, society of petroleum engineers. SPE 144326-MS. The Woodlands, TX: Society of Petroleum Engineers. <www.onepetro.org>; 2011 [accessed 17.07.16].

2. Brock TD. Life at high temperatures. Evolutionary, ecological, and biochemical significance of organisms living in hot springs. *Science* 1967;**158**(3804):1012–19.

3. Mullis KB, Faloona FA. Specific synthesis of DNA in vitro via a polymerase-catalyzed chain reaction. *Methods Enzymol* 1987;**155**:335–50.

4. Takai K, Komatsu T, Inagaki F, Horikoshi K. Distribution of archaea in a black smoker chimney structure. *Appl Environ Microbiol* 2001;**67**(8):3618–29.

5. Beatty JT, Overmann J, Lince MT, Manske AK, Lang AS, Blankenship RE, et al. An obligately photosynthetic bacterial anaerobe from a deep-sea hydrothermal vent. *Proc Natl Acad Sci* 2005;**102**(26):9306–10.

6. Dorn RI, Oberlander TM. Microbial origin of desert varnish. *Science* 1981;**213**(4513):1245–7.

7. Parro V, de Diego-Castilla G, Moreno-Paz M, Blanco Y, Cruz-Gil P, Rodríguez-Manfredi JA, et al. A microbial oasis in the hypersaline Atacama subsurface discovered by a life detector chip: implications for the search for life on Mars. *Astrobiology* 2011;**11**(10):969–96.

8. Atlas RM, Hazen TC. Oil biodegradation and bioremediation: a tale of the two worst spills in U.S. history. *Environ Sci Technol* 2011;**45**(16):6709–15.

9. Hazen TC, Dubinsky EA, DeSantis TZ, Andersen GL, Piceno YM, Singh N, et al. Deep-sea oil plume enriches indigenous oil-degrading bacteria. *Science* 2010;**330**(6001):204–8.

10. Dubinsky EA, Conrad ME, Chakraborty R, Bill M, Borglin SE, Hollibaugh JT, et al. Succession of hydrocarbon-degrading bacteria in the aftermath of the Deepwater Horizon oil spill in the Gulf of Mexico. *Environ Sci Technol* 2013;**47**(19):10860−77.

11. Valentine DL, Kessler JD, Redmond MC, Mendes SD, Heintz MB, Farwell C, et al. Propane respiration jump-starts microbial response to a deep oil spill. *Science* 2010;**330**(6001):208−11.

12. Mason OU, Scott NM, Gonzalez A, Robbins-Pianka A, Bælum J, Kimbrel J, et al. Metagenomics reveals sediment microbial community response to Deepwater Horizon oil spill. *ISME J* 2014;**8**(7):1464−75.

13. Science Daily. Deepwater oil plume in Gulf degraded by microbes, study shows. <www.sciencedaily.com/releases/2010/08/100824132349.htm>; 2010; August 25.

14. Head IM, Jones DM, Larter SR. Biological activity in the deep subsurface and the origin of heavy oil. *Nature* 2003;**426**(6964):344−52.

15. Kanaly RA, Harayama S. Advances in the field of high-molecular-weight polycyclic aromatic hydrocarbon biodegradation by bacteria. *Microb Biotechnol* 2010;**3**(2):136−64.

16. Fierer N, Lauber CL, Zhou N, McDonald D, Costello EK, Knight R. Forensic identification using skin bacterial communities. *Proc Natl Acad Sci* 2010;**107**(14):6477−81.

17. Lax S, Hampton-Marcell JT, Gibbons SM, Colares GB, Smith D, Eisen JA, et al. Forensic analysis of the microbiome of phones and shoes. *Microbiome* 2015;**3**:21.

18. Lax S, Smith DP, Hampton-Marcell J, Owens SM, Handley KM, Scott NM, et al. Longitudinal analysis of microbial interaction between humans and the indoor environment. *Science* 2014;**345**(6200):1048−52.

19. Franzosa EA, Huang K, Meadow JF, Gevers D, Lemon KP, Bohannan BJ, et al. Identifying personal microbiomes using metagenomic codes. *Proc Natl Acad Sci* 2015;**112**(22):E2930−8.

20. Kotlar H, Rueslatten H, Ramstad MV, Brakstad OG. Methods detecting, characterizing and monitoring hydrocarbon reservoirs US 20060154306 A1. <http://www.google.com/patents/US20060154306>; 2006. [accessed 09.07.16].

21. Society of Petroleum Engineers. In-Situ Molecular Manipulation. <http://www.spe.org/industry/in-situ-molecular-manipulation.php>; 2010. [accessed 17.07.16].

22. Kotlar HK. Can bacteria rescue the oil industry? The Scientist. 2009; February.

23. Jones DM, Head IM, Gray ND, Adams JJ, Rowan AK, Aitken CM, et al. Crude-oil biodegradation via methanogenesis in subsurface petroleum reservoirs. *Nature* 2008;**451**:176−81.

24. Vinegar H. Shell's In-situ Conversion Process. <http://ceri-mines.org/documents/R05a-HaroldVinegar.pdf>; 2006. [accessed 27.06.16].

25. Muyzer G, de Waal EC, Uitterlinden AG. Profiling of complex microbial populations by denaturing gradient gel electrophoresis analysis of polymerase chain reaction-amplified genes coding for 16S rRNA. *Appl Environ Microbiol* 1993;**59**(3):695−700.

26. Gilbert JA, Jansson JK, Knight R. The Earth Microbiome Project: successes and aspirations. *BMC Biol* 2014;**12**:69.

27. Rideout JR, He Y, Navas-Molina JA, Walters WA, Ursell LK, Gibbons SM, et al. Subsampled open-reference clustering creates consistent, comprehensive OTU definitions and scales to billions of sequences. *Peer J* 2014;**2**:e545.

28. Caporaso JG, Kuczynski J, Stombaugh J, Bittinger K, Bushman FD, Costello EK, et al. QIIME allows analysis of high-throughput community sequencing data. *Nat Methods* 2010;**7**(5):335−6.

29. Lozupone C, Knight R. UniFrac: a new phylogenetic method for comparing microbial communities. *Appl Environ Microbiol* 2005;**71**(12):8228−35.

30. Navas-Molina JA, Peralta-Sánchez JM, González A, McMurdie PJ, Vázquez-Baeza Y, Xu Z, et al. Advancing our understanding of the human microbiome using QIIME. *Methods Enzymol* 2013;**531**:371−444.

31. Mandal S, Van Treuren W, White RA, Eggesbø M, Knight R, Peddada SD. Analysis of composition of microbiomes: a novel method for studying microbial composition. *Microb Ecol Health Dis* 2015;**26**:27663.

32. Knights D, Kuczynski J, Charlson ES, Zaneveld J, Mozer MC, Collman RG, et al. Bayesian community-wide culture-independent microbial source tracking. *Nat Methods* 2011;**8** (9):761−3.

PART IV

Informing Policy

Get your facts first, then you can distort them as you please.

—Mark Twain

CHAPTER *10*

Enabling Regulations and Empowering the Public

Shale oil and gas development in the United States has proceeded at a rate that is unusual for a relatively staid industry. As a result, in many instances, progress has outdistanced necessary regulation. This was particularly evident in Pennsylvania, the state with the first recorded oil well, by "Colonel" Drake in 1858 in Titusville, but with no recent history of hydrocarbon production. Mistakes were made, with concomitant loss of public confidence. In a scant few years, regulations were crafted that are proving very effective, but the years with limited regulation took their toll among all but academic geochemists.

One of us (VR) was a member, and for a time chairman, of the North Carolina Mining and Energy Commission (NC MEC), which was charged by the North Carolina General Assembly to produce a rule set prior to any commercial activity. This is good policy. At a recent conference on the globalization of shale oil and gas, a member of the staff of the Prime Minister of the United Kingdom expressed a similar intent for that country. The need for more effective analytical techniques kept surfacing during the deliberations of the NC MEC. The availability and cost of measurement techniques are determinants of the feasibility of rules pertaining to the measurements. The cost aspect was brought into focus by a US Supreme Court ruling: The US Environmental Protection Agency (EPA) "must consider cost— including cost of compliance—before deciding whether regulation is appropriate and necessary."[1] Even absent this ruling, reason dictates that regulations ought not to be so burdensome as to stifle the regulated enterprise. If ready technology is found wanting, delayed enforcement dates make sense, especially if they are coordinated with federally funded innovation programs. One such instance is the US Department of Energy MONITOR (Methane Observation Networks with Innovative Technology to Obtain Reductions) program described in Chapter 3, which is directed to new methods for detection of fugitive

Sustainable Shale Oil and Gas. DOI: http://dx.doi.org/10.1016/B978-0-12-810389-0.00010-3

methane emissions on rigs and tied in principle to EPA regulations now in development on curbing fugitive emissions.

WATER-RELATED REGULATION

Chapter 5 is replete with studies that inform water-related regulation. A common thread is that certain species are increasingly viewed as key indicators of fluids from oil and gas wells. A comprehensive suite of analytes makes sense for baseline measurements prior to any drilling operations. However, the follow-on testing at 6 months, 1 year, and so forth may adequately focus on sentinel species and methods. Regarding the species, it appears that total dissolved solids and chloride ions and divalent cations are fairly reliable sentinels,[2] and they are measured at relatively low cost and difficulty. Their choice was based on potential mobility in groundwater systems and the correlation with other more harmful constituents that may be more difficult to detect. Radium, for example, is correlated with total dissolved solids and, at least in the Marcellus, strongly correlated with barium. The radium:barium ratio is often constant in unconfined aquifers.[3] Incidentally, the Marcellus has higher radioactive species on average than any other shale reservoir in the United States. Strontium per se is a strong indicator of flowback water, but especially the isotope ratio Sr^{87}:Sr^{86} is considered to be a fingerprint of Marcellus waters.[4]

From a regulatory standpoint, some certainty about sentinel species would be valuable, especially for avoiding false negatives. That would make most state oil and gas commissions more comfortable about the use of sentinels in the manner described previously. In the NC MEC, there was much debate on this point at the time of crafting the regulations (tongues firmly planted in collective cheeks, we referred to the sentinel species as the canary list, after the old practice of using canaries in coal mines as sensors of methane). Since then, more studies have solidified the science, and regulators in all states would benefit from a definitive work in this space. Ultimately, commission work product has to be passed by the legislature, and any diminution in cost of regulatory compliance without loss of effectiveness would make passage more likely. Because these are usually up/down votes, every detail has to be properly crafted.

Many of the studies, possibly beginning with Osborn et al.[5] have informed an important metric: the radial distance from the rig site within which mandatory baseline and follow-up testing of well water must be conducted. Pennsylvania settled on a figure of 2500 feet and North Carolina went a trifle further with half a mile. In the latter case, not only was testing mandatory but also the operator was presumptively liable for a detected contamination.[6] This clause places an even greater burden on the efficacy of analytical methods. A technique still in the early stages of development uses strands of DNA as tracers. Unique strands are inserted into the fracturing fluid and, if found in water wells, can be traced back precisely to the oil or gas well in question. The practical value of this method is significantly enabled by the plummeting price of polymerase chain reaction (PCR) boxes (<US$10,000) despite the fact that the science is relatively new— awarded the Nobel Prize in 1993. PCR can take a single short fragment of DNA and multiply it by several orders of magnitude. The boxes are portable, and well water testing should be simple and inexpensive. Complications include ensuring survival of the fragments in high salinity and the high temperature and pressure encountered downhole. If successful, this will be a fine example of an analytical technique enabling legislation.

As noted in Chapter 5, methane intrusion into aquifers can be from natural or anthropogenic sources. No evidence supports the notion that methane intrusion is a sentinel for liquid contamination. Regulators should recognize this conclusion and treat the fluids separately. Also, the high likelihood of natural sources for methane in well water particularly highlights the need for baseline measurements prior to commencement of drilling. Research has demonstrated that the existence of ethane and larger molecules is a signature of gas from thermogenic sources, as opposed to largely terrestrial production of biogenic methane. These species should be included in the testing suite. Some sophisticated isotopic investigation methods probably do not belong in regulation, but they should be known to the state departments of environmental protection in case difficult problems arise.

Considerable evidence has been presented for the unlikelihood of fractures propagating from the reservoir up to freshwater aquifers. The furthest reach of routine fractures has been 1000 feet, with a few excursions to 1500 feet (see Chapter 5) and the solitary observation in the US Department of Energy's National Energy Technology Laboratory study of a fracture that traveled 1800 feet. Freshwater aquifers rarely are down to depths greater than 1000 feet. Most reservoir rock is deeper than 5000 feet. One could conclude that the risk is low. However, regulators must anticipate the possibility of shallower deposits being exploited in the future, such as the Triassic in the eastern United States. In states where such possibilities exist, regulators should consider requiring a minimum vertical spacing from the bottom of the deepest fresh aquifer to the top of the shallowest petroleum reservoir being exploited. The evidence suggests that spacing of 2000 feet is reasonable. We are not aware of any state with such regulation.

Underground Injection Control Class II wastewater injection wells have been responsible for earthquakes. This is no longer in dispute.[7] Also not in dispute are the potential remedies: seismic investigation of the subsurface to detect active faults and avoidance of placing wells proximal to them. Another method would be to actively monitor the injection with microseismic methods, at least until safe parameters are developed. These wells all fall under the jurisdiction of the EPA in their design, and usually the state department of environmental protection oversees the execution. Explicit regulations are necessary at the federal or state level to mitigate this hazard.

THE CHEMICALS DISCLOSURE ISSUE

Disclosure of chemicals used in fracturing fluid remains a controversial issue. Most state regulations require the full disclosure of all chemical constituents, with a trade secret exclusion.[6] This exclusion allows companies to withhold information if it constitutes intellectual property, which in this context falls into two categories: patents and trade secrets. Whether an innovation is patented or not is a legal and business decision. If a patent is filed, US law requires that the entire application be published 18 months after filing, regardless of whether the patent is subsequently granted or not. Consequently, the secret information is in the public domain in that time frame, which is a relatively short period for most innovations to be commercialized. As a practical matter, therefore, the public will be aware of the details. If the company chooses to hold the innovation as a trade secret, that is its right in any industry. The formula for Coca–Cola is the celebrated example of a trade secret. Compulsion to reveal a trade secret for any particular industry sector would not be good policy.

The North Carolina regulations provide means by which this seeming right is not improperly exercised. The rules do not allow the granting of trade secret status simply because it is sought. A special part of the MEC (now called the Oil and Gas Commission) will assess the merit of the claim in a simple procedure. A key provision is that the company seeking it must demonstrate that the information is accorded the same internal protections against public disclosure as any other of its trade secrets of value. Such a verification step is likely to ensure that only legitimate claims are brought forward. As a matter of detail, secret status is always sought by the service companies, not oil and gas operators.

In most instances, the matters that are considered trade secrets are the recipes, not the chemicals. This knowledge is perceived as a competitive advantage. The actual proportions in the fracturing fluid going in are not those of the fluid returning. That is because some chemicals are consumed and others are absorbed in the formation or return in a different form. Although it is important to know what chemical species are going in, in order to test for them in suspected contamination situations, the actual *proportions* are not particularly important. Furthermore, most recent regulations, including those in North Carolina, do not permit benzene, toluene, ethylbenzene, and xylene (BTEX) or other aromatic constituents to be in the formulation. When these chemicals are present in the flowback water, they are sourced from the rock below. The same is true for radioactive species. In Chapter 5 it was noted that the likelihood of groundwater contamination is far greater from surface spills than from wells leaking return fluid. In these instances, again, the chemical description, rather than the proportions in the fracturing fluid, is important.

AIR-RELATED REGULATION

The most recent issues concerning air pollution associated with oil and gas activities have been fugitive methane emissions and health outcomes from volatile organic chemicals. The latter have been studied less, although the state of New York declared a moratorium on the industry premised on perceived dangers to health. Both of these areas have been constrained by the maxim, "That which you cannot measure, you cannot regulate or otherwise control." As described in Chapter 3 the Environmental Defense Fund has taken a lead in sponsoring and funding four new sensor developments for detection of fugitive methane. The US Department of Energy has likewise, through its ARPA-E arm, funded 11 developments on a fast track. In many of these, there is no new measurement science; they target the issues of cost and ease of deployment. Consequently, but for the attention to cost driven by the Supreme Court decision mentioned previously, the EPA should be in a position to set regulatory limits, and as of mid-2016 it was in the process of doing so. This is timely because fugitive methane has rapidly become the most watched environmental issue. Although there is general acknowledgment of the existence of these emissions, the disputes center on the amounts and, in one case, the allegedly flawed measurement methods.[8,9] The claimed flaw is once again not in the measurement science but in the method of execution, particularly the proper switchover of the device from low-volume to high-volume detection. There is unanimity on one point: Emissions need to be curbed.

Particulate matter regulation is sparse. Particulate matter consists of airborne dust, such as from proppant handling, and carbon from combustion of fuel. The former is in the jurisdiction of the Occupational Safety and Health Administration. The latter is more difficult to measure and quantify. Currently, measurement of PM10 and PM2.5, respectively particles smaller than 10 and 2.5 μm, is done at stations positioned throughout the counties. The locations and frequencies of measurement are generally not suited to the practicalities of shale oil and gas development. The particulate matter is from diesel operations on rigs and truck traffic. The latter is highly episodic because trucks deliver water or sand in batches. Consequently, regulating this feature will be complex, in part because much of the effect will be experienced not on the rig site but in the surrounding community along the drive path. The MicroPEM mentioned in Chapter 4 may well be an innovation that facilitates measurement. State regulators should require that the truck convoy paths be disclosed to the regulatory authority a certain number of days prior to the event. This information can be used to fairly quickly deploy units of MicroPEM or another type of sensor to locations on the route. Similarly, populations on the route could be narrowed down and the most exposed portion of the cohort could be identified for possible personal monitoring. If the truck routes could be permanently retained, perhaps as an element in the drilling permit application, then permanent stations on the route could be maintained.

A regulatory aspect of the truck traffic caused by oil and gas operations is the need for schemes to pay for road and bridge damage. Most rural roads and bridges are not intended for heavy traffic, certainly not for bursts of heavily loaded vehicles. An impact fee, distinct from severance tax at the state level, is merited. Severance taxes usually accrue directly to the state and are proportional to the total production. Impact fees should accrue largely to the affected communities and be proportional to the amount of effort, because effort is proportional to truck volume. Truck volume and frequency will be dictated by the design of the well pads. Simplistically, the volume is proportional to the number of fracturing stages. Each stage uses approximately the same amount of water and sand. The frequency will depend on whether the operation is of conventional design or uses batch drilling. Furthermore, reuse of flowback water should be rewarded because it directly reduces freshwater use. In many instances, this water is brought in by truck. Reusing wastewater reduces traffic related to wastewater transportation for disposal.

Regulations on air emissions sometimes run afoul of separation of jurisdictions. When the NC MEC issued draft oil and gas rules for public comment, one of the top three issues raised was that the MEC needed to write rules on air emissions. The MEC was statutorily prevented from doing so because air emissions from all industries were in the purview of the state's Environmental Management Commission.

Cost-effective measurement of volatile organics is needed to inform regulation. Much of the evidence for these airborne species is in the form of health outcomes, often on an anecdotal basis. In early 2016, the EPA issued draft regulations for benzene measurement on refinery perimeters.[10] Of the BTEX compounds, only benzene was cited, possibly because of the difficulty in measurement of multiple species. It could also be that benzene was viewed as a sentinel compound. If the innovation described in Chapter 4 comes to fruition, more stringent regulation could be emboldened. In particular, measurement on the perimeters of oil and gas production sites could become feasible. This may be necessary to inform the public and, by extension, the regulators, especially in moratorium jurisdictions such as New York and France.

FLARING REGULATION

Flaring is regulated at both federal and state levels. The federal government oversees rules that apply to refineries and chemical plants, and separately to rig sites. In general, regulation is deferred to the states except for federally owned land on which there is production. In the spring of 2016, the EPA gave a clear signal that it was close to formulating a rule set directed to flaring, based on the latest figures in the national inventory. There is a general expectation of target reductions in the vicinity of 50% over 10 years.

For several years, North Dakota has implemented its own stair-step regulation set, as shown in Fig. 10.1. The pattern was intended to allow time for remedial action. The decrease to 15% was extended by 10 months to late 2016 in recognition of the downturn in the industry. The average flaring rate has fallen below target levels, but curiously the cumulative figure increased in 2015 (not shown). This likely means that the overall volume of associated gas has increased due to selective operations. Throughout the entire United States during this low oil price period, there has been a shift of production to the high-grade properties. This accounts for the fact that even with a plummet in working rigs, the actual production has not declined much. Of course, this must be measured against the trend in previous years of the decade, when the country added 1 million barrels per day each year. Regarding the fact that the quantity of flared gas increased in North Dakota during approximately the past 2 years, even while gains were made in the percentage of gas flared, the cause could well be the high grading. In the end, what matters is the total gas flared, not the percentage.

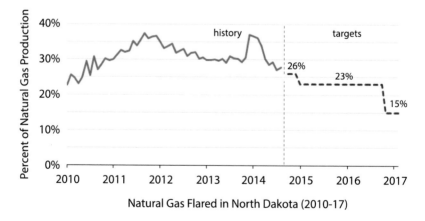

Figure 10.1 North Dakota flaring history and new regulations and actual performance. Source: Energy Information Agency.

One takeaway from the foregoing discussion is that the statistics on typical flared gas per pad, as shown in Fig. 2.6, may now be skewed to higher numbers. This could well be temporary, but it still constitutes a guide to development of amelioration methods, which may now be sized larger. The ideal mode is a size capable of handling 50,000 standard cubic feet per day that can be designed in modular fashion in order to be doubled when needed.

The incentive to build gas pipelines is reduced by the uncertainty in oil prices. This uncertainty, together with current projections[11] of low prices for several years, will discourage the construction of new export pipelines for associated natural gas. Further exacerbating this trend will be the currently predicted low natural gas prices for the same period. On the other hand, low oil prices have dramatically reduced new well construction. Prior to this situation, the issue faced by regulators in North Dakota and other producing states was that restrictions on flaring could reduce overall activity and hence revenue for the state. The stair-step regulation was an effort at compromise. This concern is scarcely in play now in mid-2016 because production is declining for economic reasons.

PUBLIC'S RIGHT TO KNOW AND CITIZEN SCIENCE

The public's right to know is almost always cast in terms of an understanding of the risks to its general well-being. One could argue that people should also be made aware of some of the collateral benefits of shale oil and gas. However, as a practical matter, well-being issues are much more up close and personal than the effect on national security or even greenhouse gas mitigation through shale gas substitution for coal in power plants. Of the possible concerns, the two major ones are well water contamination and airborne chemicals.

The more advanced regulations require baseline testing of water sources within a radius of at least 2000 feet from a drilling operation. The suite of analytes is comprehensive. Follow-on testing after commencement of operations is required for either the full suite or a sentinel list.[6] Time spans between testing are at least 6 months. If a citizen has reason for concern, low-cost access to testing would provide assurance. If the unique DNA strand tracer approach mentioned earlier in this chapter is perfected, the test to detect is inexpensive. The necessary

PCR box, at less than US$10,000 a copy, could be made available from an extension service. Although this would eliminate or confirm direct well contamination from well leaks or wastewater impoundments, it would not serve the purpose for spills of neat chemicals at the surface.[12] That might require wet chemistry.

Volatile organic chemicals is untrodden territory in comparison. Regulations for perimeter testing of benzene are only just being promulgated in 2016, and only for refineries. None appear on the horizon for perimeter testing of drilling units. Advances in analytical methods could embolden legislation in this space. If the innovation described in Chapter 4 is successful for the price point targeted, then devices could become available in communities for spot testing for citizen science or simply peace of mind. Citizen science is a burgeoning endeavor and involves ordinary citizens making measurements when concerned by signals such as odor or malaise that could be ascribed to volatile chemicals. Such efforts are generally hampered by poor methodology or the absence of organized training required to make correct measurements using EPA-approved procedures.

Citizen science would be significantly aided if analytical methods agnostic to the skill of the operator were developed. All of the MONITOR program awardees are required to have data transmitted automatically and wirelessly to a central location, a feature that not only reduces the burden for the citizen operator to deliver results but also makes the results not susceptible to manipulation. Research should require this feature if citizen science is a target market. Ideally, even air sampling would be automated, leaving little margin for error. Monitoring of particulate matter is now completely feasible in small wearable devices, as described in Chapter 4. In all of these examples, documentation of the location of testing is a detail that must be worked out.

CONCLUSIONS

Regulation and control of a phenomenon require measurement. Policy takes many forms beyond simple regulation, but in most cases, this axiom holds. Other countries are poised to follow the example of the United States with regard to shale oil and gas. Argentina may be the best situated with respect to quality of the resource.[13] A recent conference at the University of North Carolina titled "Global Frac'ing, What Has to Change for It to Be a Game Changer?" concluded that "It is not only about the rock." That could well serve as the conference report title. The point is that without quality rock and the ability to exploit it, nothing happens. However, this is a necessary but not sufficient condition. Sufficiency lies in myriad enablers, including fiscal regimes, environmental regimes, availability of service supply, and, importantly, ensuring societal comfort with the enterprise.

The new areas could begin with studying the regulations in the US states that have recently issued them. Responsible authorities should follow advances in measurement techniques to enable regulation that would otherwise be less feasible or fair. Fugitive methane is a case in point of investigation yielding an unexpected conclusion—that midstream and downstream architecture was responsible for the bulk of the fugitive emissions. The point holds irrespective of whether the natural gas is from shale or conventional sources, and it implies that existing infrastructure needs attention even in the absence of shale oil and gas exploitation. If they follow the space closely, one or more countries may well take the lead in regulations enabled by new measurement capability. Candidates for such leapfrogs would be the monitoring and control of volatile species and particulate matter proximal to oil and gas activity.

Every form of energy has a risk:reward profile that determines desirability. In each case, sustainable production must be the goal. All three legs of the sustainability stool are equally important. Without profit there is no enterprise, and yet this cannot be at the expense of the environment or the well-being of those in direct contact with the enterprise. The first two legs are heavily dependent on measurement science and form the basis for this work. The assessment and amelioration of societal impact also are enabled by measurement science of a type that is not the subject of this book. That which cannot be measured, cannot be regulated or otherwise controlled or exploited.

REFERENCES

1. Supreme Court Decision. Michigan et al. versus Environmental Protection Agency et al. <http://www.supremecourt.gov/opinions/14pdf/14-46_10n2.pdf>; 2015. [accessed 22.04.16].

2. Coleman PC. *Produced formation water sample results from shale plays.* 2012. <https://www.epa.gov/sites/production/files/documents/producedformationwatersampleresultsfromshale-plays.pdf>. [accessed 24.06.16].

3. Zhang T, Gregory K, Hammack RW, Vidic RD. Co-precipitation of radium with barium and strontium sulfate and its impact on the fate of radium during treatment of produced water from unconventional gas extraction. *Environ Sci Technol* 2014;**48**(8):4596–603.

4. Warner NR, Jackson RB, Darrah TH, Osborn SG, Down A, Zhao K, et al. Geochemical evidence for possible natural migration of Marcellus formation brine to shallow aquifers in Pennsylvania. *Proc Natl Acad Sci* 2012;**109**(30):11961–6.

5. Osborn SG, Vengosh A, Warner NR, Jackson RB. Methane contamination of drinking water accompanying gas-well drilling and hydraulic fracturing. *Proc Natl Acad Sci* 2011;**108**:8172–6.

6. North Carolina Regulations. Oil and Gas Conservation. <http://reports.oah.state.nc.us/ncac/title%2015a%20-%20environmental%20quality/chapter%2005%20-%20mining%20-%20mineral%20resources/subchapter%20h/subchapter%20h%20rules.pdf>; 2015 [accessed 06.30.16]. [See sections 15A NCAC 05H. 1611 (well installation, including testing cement integrity), 15A NCAC 05H. 1609 (surface casing standards) and 15A NCAC 05H. 1504 (pits and tanks)].

7. Walsh III RF, Zoback MD. Oklahoma's recent earthquakes and saltwater disposal. *Sci Adv* 2015;**1**(5):e1500195.

8. Brownstein M. *The UT methane studies: critique and response* 2015. <http://blogs.edf.org/energyexchange/2015/03/03/the-ut-methane-studies-critique-and-response/>. [accessed 19.06.16].

9. Brownstein M, Hamburg S. *Keeping an important methane research question in proper perspective* 2016. <http://blogs.edf.org/energyexchange/2016/06/09/keeping-an-important-methane-research-question-in-proper-perspective/>. [accessed 19.06.16].

10. US Environmental Protection Agency. [Proposed refinery rules]. <https://www.gpo.gov/fdsys/pkg/FR-2016-02-09/pdf/2016-02306.pdf>. 2016 [accessed 24.06.16].

11. US Energy Information Administration. *Short Term Energy Outlook* [July 12, 2016]. <https://www.eia.gov/forecasts/steo/pdf/steo_full.pdf>; 2016. [accessed 16.07.16].

12. Drollette BD, Hoelzer K, Warner NR, Darrah TH, Karatum O, O'Connor MP, et al. Elevated levels of diesel range organic compounds in groundwater near Marcellus gas operations are derived from surface activities. *Proc Natl Acad Sci* 2015;**112**(43):13184–9.

13. Rao V. *Shale oil and gas: the promise and the peril.*. 2nd ed. Research Triangle Park, NC: RTI Press; 2015 [Chapter 28].

INDEX

Note: Page numbers followed by "*b*," "*f*," and "*t*" refer to boxes, figures, and tables, respectively.

Printed in the United States
By Bookmasters